cross-pollinations

THE *CREDO* SERIES

A *credo* is a statement of belief, an assertion of deep conviction. The *Credo* series offers contemporary American writers whose work emphasizes the natural world and the human community the opportunity to discuss their essential goals, concerns, and practices. Each volume presents an individual writer's *credo,* his or her investigation of what it means to write about human experience and society in the context of the more-than-human world, as well as a biographical profile and selected bibliography of the author's published work. The *Credo* series offers some of our best writers an opportunity to speak to the fluid and subtle issues of rapidly changing technology, social structure, and environmental conditions.

cross-pollinations

THE MARRIAGE OF SCIENCE AND POETRY

GARY PAUL NABHAN

Scott Slovic, *Credo* Series Editor

Credo

MILKWEED EDITIONS

Published 2004 by Milkweed Editions
Printed in Canada
Cover art, "Zinnia & Dragonfly," iris print by Anna Tomczak
The text of this book is set in Stone Serif.
04 05 06 07 08 5 4 3 2 1
First Edition

"Urn-Burial and the Butterfly Migration," from *The Collected Poems of Amy Clampitt* by Amy Clampitt. Copyright © 1997 by the Estate of Amy Clampitt. Reprinted with permission from Alfred A. Knopf, a division of Random House, Inc.

"Requiem for Sonora," from *Of All the Dirty Words* by Richard Shelton. Copyright © 1972 by Richard Shelton. Reprinted with permission from the University of Pittsburgh Press.

Special underwriting for this book was generously provided by The Toro Foundation.

Milkweed Editions, a nonprofit publisher, gratefully acknowledges support from Emilie and Henry Buchwald; Bush Foundation; Cargill Foundation; DeL Corazón Family Fund; Dougherty Family Foundation; Ecolab Foundation; Joe B. Foster Family Foundation; General Mills Foundation; Jerome Foundation; Kathleen Jones; Constance B. Kunin; D. K. Light; Chris and Ann Malecek; McKnight Foundation; a grant provided by the Minnesota State Arts Board through an appropriation by the Minnesota State Legislature, and a grant from the Wells Fargo Foundation Minnesota, and a grant from the National Endowment for the Arts; Sheila C. Morgan; Laura Jane Musser Fund; National Endowment for the Arts; Navarre Corporation; Kate and Stuart Nielsen; Outagamie Charitable Foundation; Quest Foundation; Debbie Reynolds; St. Paul Companies, Inc.; Ellen and Sheldon Sturgis; Surdna Foundation; Target Foundation; Gertrude Sexton Thompson Charitable Trust; James R. Thorpe Foundation; Toro Foundation; and Xcel Energy Foundation.

Library of Congress Cataloging-in-Publication Data

Nabhan, Gary Paul.
   Cross-pollinations : the marriage of science and poetry / Gary Paul Nabhan.— 1st ed.
     p.   cm. — (Credo)
   Includes bibliographical references.
   ISBN 1-57131-270-6 (pbk. : acid-free paper)
   1. Literature and science. I. Title. II. Credo series (Minneapolis, Minn.)
   PN55.N33 2004
   809.1—dc21

2003011419

For Alison Deming,
fellow traveler on the trails
of scientific and poetic explorations,
and sweet inspiration

cross-pollinations

by Gary Paul Nabhan

At the heart of this language is a constellation of words that includes *connectedness, conviction, courage, improvisation, inquiry, intuition, introspection, passion, pressure,* and *process.*

—Michael Brenson, *Visionaries and Outcasts*

cross-pollinations

cross-pollinations

by **Gary Paul Nabhan**

CROSS-POLLINATIONS
An Overture

> *The true artist, like the true scientist, is a researcher*
> *using materials and techniques to dig into the truth*
> *and meaning of the world in which he himself lives,*
> *and what he creates, or better perhaps, what he brings*
> *back are the objective results of his explorations.*

—Paul Strand

The writer's life is much like that of a sphinx moth,
which appears out of the darkness to hover suddenly
above a freshly opened blossom, coming under the spell
of its pungent perfumes. The moth lingers before the
illuminated flower for a moment, then dips into the
ephemeral world hidden within the floral tube, where
it draws energy from the flower's nectars and perhaps
dreams from its alkaloids. Another moment passes and
the moth is nowhere to be seen. It is loading up with the
pollen, which it will transport from flower to flower, en-
abling something potentially far more lasting to occur—
cross-fertilization and regeneration. As for the moth itself,

3 ☞

it has already darted off through the darkness, seeking another floral encounter that might nourish its own life and open new possibilities for others as well.

The individual moth, at least in its metamorphosed form as a somewhat flighty floral visitor, may not live long. And yet the world we live within is filled with perfumes of passion and petals of subtle beauty thanks to its coevolutionary dance with flowering plants. These scents and scenes have somehow gained staying power in our memories, disproportionate to the period of time we have stood in their presence. So too with the work of poets, storytellers, and literary naturalists.

Most moths work the night shift, although some are inclined more to crepuscular moments, dazzled by the magic of dusks and dawns. I cluster with the latter. Like one of my writing mentors, the late poet Bill Stafford, I have typically let my pen loose to fly during the *madrugada*, the twilight time just before sunrise. It is the time of day when the telephone is least likely to ring and disrupt the flight of ink.

Only once in my life did I do much writing while the sun was high in the sky, for I've usually had a day job that's separate from my literary pursuits. However, in the early 1990s, my interest in pollination ecology took a turn toward the nocturnal, and the little time I had left for writing occurred around high noon. By late afternoon, I left my desk behind so that I could follow sphinx moths around on their night shift, tracking their movements from dusk to dawn. I was captivated by the way they engaged themselves with two late-bloomers—the sacred datura and the night-blooming cereus. These two desert plants play major roles in the dramas that follow, but before I fly too far out ahead of this particular story, leaving it behind for the scent of another, let me hover before it a moment longer.

During that period in the nineties, I preferred to live on the borderline—that is to say, to dwell on the boundary between the United States and Mexico, the interface between science and art, indigenous and exotic, tradition and risk. Just before sunset, I would set out along the border fence between the wildlands of Organ Pipe Cactus National Monument in Arizona and the floodplain fields along the Río Sonoyta of Sonora, to see how many datura and cereus flowers might bloom that night. Only in late afternoon could I detect whether they were ready to slowly spiral open, morphing from bud to blossom. If both cactus and datura were to shift into their reproductive states on the same night, seeking the companionship of sphinx moths, I too would move back and forth across the border fence, following them. That is, I would track the flow of pollen between two countries, at least until Customs closed their gate at midnight, blocking my passage. Around midnight, I was forced to stay on one side of the border or the other, to hunker down by a single flush of flowers, to stay beneath their spell until they closed up shop an hour or so after dawn had come and gone.

The night-blooming cacti clung to the rocky ridges paralleling the border, but the bottomlands belonged to datura plants, for they thrived where tailwaters from agricultural irrigation gathered at the ends of long rows of pima cotton. Nevertheless, the two kinds of plants could be seen within a dozen yards of one another, where they benefited from the same pool of pollinators. Even

5 ☞

on the best of nights for cactus blossoms, they were out-numbered ten to one by the stunning blooms of datura. While pleasant, musky volatiles filled the air above one of the blooming cereus species, a malevolent odor associated with the rank foliage of neighboring datura sometimes overwhelmed the scent of its flowers. This was especially true whenever the leaves of daturas were bruised by livestock rubbing against them as they passed along the field's edge. The poisonous and perilous alkaloids of datura foliage seemed to leak out of the leaves and stems to imbue the very atmosphere around these plants, so much so that my O'odham Indian neighbors warned me to stay clear of these narcotic–laden weeds if I didn't want to go crazier than I already was.

If I succumbed to vertigo or insanity on those summer nights, it does not now register in my memory. What I do remember is a certain joy at having arrived within a patch of plants just as a flower bud began to reel out to catch a moth with its lure. A tightly twisted, purple-tinted bud would quickly become a shimmering trumpet, playing taps for the fallen sun, the music required for moths to dance. The moths were dazzled by these blossoms' iridescence just as they are by the brightness of a back porch light; just as writers are by the brilliance of a novel storyline. But what truly called the moths in was not a sound, it was an olfactory sensation, an arousing fragrance capable of making any hawkmoth within smelling distance appear, then genuflect, as if in raptured prayer.

I diligently wrote field notes after every search I made for night-hawking sphingids and their host plants, but I also learned what I could from scientists and herbalists of various traditions who were as enamored as I was with the matching of moth and blossom. For instance, insect neurobiologist Rob Raguso helped me understand how moths find the nectar they need. The male moths are alerted to their presence by a particularly strong scent flume that wafts through the evening air several yards above the stony ground, but they merely pass by unless there is a visual cue as well. Fortunately, the pale blossoms of datura and cereus stand out against the darker hues of ironwood and creosote that often surround them. When the moths catch a whiff of this invisible trail of volatile aromatic oils, they flutter to where it is concentrated, then follow visual cues until they hover before the funnel-like flower, which is filled with a pool of nutritious nectar. As Rob Raguso discovered, only when the male moths reach the spot where the night air is most intensely saturated with fragrances do they switch from olfactory cues to visual ones, extending their proboscises and sampling the nectar within the blossom.

Perhaps that is the way we are least like moths. Moths are led by their noses most of the time, so that their synesthetic mix shifts toward the visual only in the final stages of floral matchmaking. In contrast, humans can be excited by the merest glimpse of a potential partner from afar, but that partner's pheromones hardly register with us until we are within arm's reach.

Once, when I had dusted the insides of an open blossom with an ultraviolet dye powder, which the moths transported from bloom to bloom, I discovered something neurobiologists wished to know, something that had been outside their line of focus. I discovered that a *Manduca* hawkmoth that first visited a dye-dusted flower was just as likely to fly 350 yards before dipping into another corolla as it was to move to the closest plant in bloom. They are such strong fliers that they seem unimpressed by optimal foraging theories, which predict that they should move in the most energetically efficient manner, consuming more nectar by moving the shortest possible distance between two points.

And so it is with poets. It is largely beyond their capacity to predict what the next item will be to capture their imaginations (to feed their souls), or how far they must be willing to travel before reaching their goals.

By the time midnight had come and gone, this monitoring of moth flights and nectar pools had left me fully satiated. The season ran from late June to early September, with tomato hornworms and white-lined caterpillars transformed to hawkmoths in several pulses over that period of time. I would also track the plants as best as I could, following their own metamorphoses from bud to bloom to fruit to dispersed seed.

Oddly enough, I did not initially recognize that period as one in which I was particularly creative or productive. In fact, while actively engaged in such fieldwork I often had the sinking feeling that my writing (other than my scribbling of field notes) was being profoundly crippled by the fatigue and disrupted circadian rhythms from which I suffered. I simply couldn't imagine at that time what working on the night shift would bring to me over the long run. Nevertheless, those patches of night-blooming plants on the borderline somehow became the nursery grounds for three books, two essays of creative nonfiction, and a half-dozen scientific articles on mutualistic ecological relationships. The discussions I had with colleagues in the field became the initial impetus for protecting more ironwood forest habitats where cactus–moth interactions remained intact. Such protection was later assured by the enthusiasm for desert habitats demonstrated by Secretary of Interior Bruce Babbitt. Fortunately, Secretary Babbitt urged President Clinton to establish a 120,000 acre Ironwood Forest National Monument by presidential decree in 1999.

Only in retrospect could I recognize how that shift in activity and perception led to an outburst of poetic and scientific creativity. If I could attribute that creative outburst to any particular factors, it would be these:

- I felt free to move from flower to flower, back and forth between blooms of both cereus and sacred datura, and through the process of emulating the movement of moths, the metaphors of cross-pollination began to guide my *modus operandi.*

- I felt free to move between the practices of field science and the literary arts without being hampered by disciplinary boundaries.

- I felt free to move between cultural landscapes (with their anthropogenic attractions) and wild habitats that were hardly populated by humans (but amply populated by other organisms).

- I felt free to spend lengthy periods of time alone, in silence, and these were intermixed with other periods of intense and playful exchange with field

ecologists, neurobiologists, poets, plant physi-
ologists, rangers, and photographers. We shared
images and tentative facts freely among us.

- I felt free to move between two countries, four
  cultures, and five languages (Spanish, English,
  O'odham, Seri, and scientific Latin). Not only
  did I take in information from all these lan-
  guages and cultural perspectives on a regular
  basis, but I freely borrowed metaphors and syn-
  tactical constructions from them as well.

And so cross-pollination became more than a topic of
study; it also became the process by which I worked. As
I explain in a poem called "Every Night":

> Every night wild creatures
> fill up the heavens, some taking the form
> of those constellations you know so well:
> corvus, scorpio, ursus—others
> more like hawkmoths and long-nosed bats
> which whisk themselves off into the darkness
> seeking the sweetest of evening nectars
> willingly diving for frenzies of feeding.
>
> On every night we break loose
> something at least this miraculous
> a moment mattering just this much
> waits for us, is stalking us, within our immediate
>   world.
> It's all here before us, every night.
> The beasts of the skies swirling around us,
> the animals chanting, the beckoning sea.
> Feel its pull there upon us? Are you ready to dive?
> Is that your hand? Every night?

As I reflect back on the course of my formal education, it is amusing to recall how many teachers admonished me to avoid such interdisciplinary pursuits:

"Stick to one subject," they said.

"Ground yourself deeply in the skills of one discipline," they warned, "or else your intellect will be frittered away. You may think you're a jack of all trades, but you'll be the master of none."

"If you squander all your time reading poetry and novels," one mathematics teacher admonished me, "you'll never be able to master the rigors of science and math, which are the most precise ways of understanding how the world works."

I heard it coming from the other direction as well. "Your poetry will become even more unintelligible if you continue to burden your free verse with the weight of scientific terms."

No wonder I became a frequently truant student, a high school dropout. Even when I was admitted to college on probation without even a GED, I lasted only a semester and a half before fleeing on a leave of absence to work at Environmental Action, the Washington, D.C., headquarters for the first Earth Day.

Fortunately, I did encounter other teachers—however few in number—who felt no immediate need to slice the world up into competing disciplines and perspectives; Karen Gunderson, Dorothy Ives, Jim Mason, Curt McCray, Jane Taylor, David Lyon, and Amadeo Rea were among them. I soon took summer writing workshops with poets such as Bill Stafford, Gary Snyder, and Pattiann Rogers, all of whom loved the quirky vocabulary of field science and embraced the vision of ecological history that regards nature as a measure of the fitness of human endeavors. Natural and cultural scientists such as Robert Euler, Howard Scott Gentry, Richard Felger, Bunny Fontana, Wes Jackson, and Stuart Struever encouraged me to use metaphor as well as technical precision in my writing. And friends such as Richard Nelson, Larry Stevens, Robert Michael Pyle, Ofelia Zepeda, Steve Trimble, Barbara Kingsolver, Carlos Martinez del Río, and Sara St. Antoine demonstrated to me by their very example that it was indeed possible to be a simultaneous participant in both of the two cultures that C. P. Snow described—the one consisting of geeky scientists, the other of spacey artist types.

Perhaps more than any other factor, my fieldwork with plants, moths, butterflies, bees, bats, and hummingbirds has helped me understand that cross-pollination is not merely a metaphor but a requisite for sustaining the diversity of life on earth. It is a survival necessity for certain organisms, the only way they are able to continue their legacy.

The downside of this understanding is that I have learned in my heart what extinction is all about, especially the extinction of relationships. I have touched with my own hands plants so rare, so threatened, that they are no longer able to attract the pollinators they need to set seed and pass on their genes. Extinction seldom comes in one fell swoop, with a bulldozer's scoop or the shot of a single gun barrel. Instead, it occurs when a web of supporting relationships unravels. It occurs whenever we or any other species are unable to sustain mutually beneficial interactions with those around us, those with whom we have been historically associated.

No, few endangered plants and animals suffer their ultimate extirpation by being physically removed from this earth. They die by suffering from the loss of ecological companionship.

Cross-pollination is not some perk or frill that benefits only an elite few. Tens of thousands of kinds of plants need cross-pollination if they are to yield fertile seeds and plump, ripe, delicious fruit. Hundreds of thousands of insects, and thousands of other (vertebrate) floral visitors are nourished by nectar and pollen. Artists and scientists also need cross-fertilization or else their isolated endeavors will atrophy, wither, or fall short of their aspirations. Linguists now suggest that cross-cultural exchanges have been among the driving forces of human evolution over the last two million years. The spark that moves between us ultimately has the capacity to sustain us over the long run. In the following stories, I will try to follow that spark.

BLENDING FIELD SCIENCE AND THE ARTS
Insights from the Cereusly Color-Blind

> *It is easy to perceive that the prodigious variety which appears both in the works of nature and in the actions of men, and which constitutes the greatest part of beauty of the universe, is owing to the multitude of different ways in which its several parts are mixed with, or placed near, each other.*

—Jakob Bernoulli, *Ars Conjectandi*

As members of my Lebanese-American clan have reminded me, I grew up with a pen, a piece of chalk, or a pencil always in hand. In fact, my hands would often be covered with ink marks and chalk dust for days at a time. Whenever I could find a place to spread out a piece of paper, I would begin to string a chain of letters across the page, making line after line of them. To explain this behavior, one of my cousins has argued that because Phoenicians invented the alphabet, many of their Lebanese descendants have a natural predilection for writing at a precocious age.

This might be stretching it a bit, at least in my case, for I was initially more interested in how letters graphically fit together than in what they could collectively mean. Still, an interest in words and their meaning did follow. My first poem was published in a local newspaper when I was still in third grade. But my artistic inclinations had been nurtured by those around me even earlier; when I was five, my teachers submitted

15 ⌦

some of my finger paintings for a competition, and one made it into the Sunday pages of the *Gary Post-Tribune*. It prompted my favorite aunt to climb up the stairs on the sand dune in front of our house that morning to praise me for my composition: a dump trunk spewing a shower of sand out its back gate. I confided in her that when I grew up, I wanted to be an "artisipaint."

"Honey, do you mean a *painter*? Or do you mean you want to be an *artist*, not just someone who paints walls or houses?"

"Artisipaint," I retorted. "Artist what paints, not just one what draws. It's cuz I like colors."

I did like to work with colors: those that oozed out of paint sticks, crayons, colored pencils, water colors, and magic markers. I continued to mess with colors even though, to my chagrin, I later learned that the hues that I saw were not necessarily those that my friends and neighbors saw. That realization was traumatic but ultimately liberating. It occurred when I was fifteen and immersed in innumerable projects involving tempera, oil, and water colors. I found myself among the many students pulled out of art class to be sent to the school nurse's office for a routine battery of perception tests. When it was my turn to sit with the nurse, she showed me a series of spheres comprised of colored dots. She asked me to tell her the numbers I saw in each of these mosaics.

I sat before her, unable to speak. Even though I diligently scanned each sphere with the hope of find-ing a number, I saw only worm- and germ- and sperm-like shapes squirming around inside the circles. In only four of the twenty-five spheres presented to me could I discern even a single numeral. For the first time in my life, I had flat-out failed a test. Analyzing which spheres had tripped me up and which I had found numbers within, the nurse announced to me that I was red-green color-blind.

"It's a disability that you should take seriously," she added, without any more explanation. Then she dismissed me so that I could return to the last ten minutes of art class.

When I rejoined the other art students, I slunk into the back row and put my head down on the desk in the corner, unable to look anyone in the eye or to paint a single object. If Dorothy Ives, my art teacher, noticed my aberrant behavior that first day, she didn't say anything to me about it. But the next afternoon, when I came into class wearing nothing but gray, she felt obliged to leave her desk and talk with me. I was wearing pale gray Levi's, a charcoal gray wool sweater, gray cotton socks; hell, I'd even spray-painted my Converse All-Stars steel gray. My head was already down on the desk again when Dorothy Ives tapped me on the shoulder.

"Something up?"

"Nope. It's all down."

"Looks like it's all gray too. You gonna keep your head down for the entire period or get your butt working on the painting you didn't finish last week?"

"I can't. And anyway, it's no use. Yesterday the nurse told me I got a handicap. Color-blindness. If I can't see red or green, how'm I ever gonna be an artist?"

Dorothy Ives gripped my arm with one hand and tapped the fingers of her other hand on the desktop. She bit her lip, then turned away from me for a moment.

17 ⇐

That's when I glanced up from where my head was still buried in my arms. All I could see was Dorothy Ives' flaming red hair, hair that had been dyed three different tints already during that single semester. "Red," I said to myself, *"red."* I mulled over a contradiction in my thinking: I saw that her locks had been dyed red this month, but the nurse had told me that I couldn't see red at all. So what was I seeing? I had no chance to answer that question, since Dorothy had suddenly turned back toward me, her face fiery red. She was gripping my arm harder than she had ever done before.

"Who in the hell told you that just because you're red-green color-blind that you cannot be an artist? Did you make that up? Did the nurse or someone else actually tell you that? If the nurse told you that, she should be tied to a stake and burned!"

Dorothy Ives was suddenly furious, joyously angry. She was filled with the same fervor that had lured me into taking her classes every semester since eighth grade, when she had intentionally given me a B+ instead of an A- to keep me from getting straight A's. ("The sooner you quit worrying about how others grade you and start concentrating on what work is satisfying to you, the better off you'll be!" she had barked at me. Then she had grinned, nudged my arm, and disappeared into the teachers' lounge.)

This time, it was clear that she was brooding over what to do next. "Gary, do you really think you can't be an artist just because you're color-blind?"

"For crap sakes, Dorothy, its just common sense, isn't it? I thought about it last night, that it's probably why none of the other students have ever liked my paintings. Have you ever noticed that my stuff must not look right to normal people?"

"Now I've warned you—warned you and the others—never to use words like *normal, average, fashionable,* or *popular* in my classroom." She was tapping on the desk-top again. "We don't take votes to determine which art is of lasting quality. You can leave this classroom if you're going to speak such vacuous nonsense."

"All right. I mean, I know what you're always telling us: *Beauty is relief from monotony.* But if no one else can stomach my mix of colors, what the heck do you want me to do?"

She bit her lip again, then whispered firmly: "Paint what *you* see, that's all. That's all I'm asking of anyone."

"Even if it gives them headaches?"

"I don't care if they get bellyaches, headaches, or sore eyes! Look, this is between you and me—I don't care what the masses think. I want to get a sense from you of what you actually see, . . ." Now I was sure she was ad-libbing, for she added, "Because, in short order, we're gonna find out what your personal aesthetic is. I mean, your personal color wheel."

I moaned. "Just how are we gonna do that?"

"You and I will become scientific experts on color blindness," she announced, "to figure out what you see—what you do and what you don't see—so that then we can use that knowledge of science to guide your art."

At that point, she'd lost me. What did science have to do with art? How could scientifically knowing more about color blindness guide my art or make my aesthetics richer? She must have seen the confused look on my face, for she hesitated a split second; I felt like we were both falling backward off a steep cliff, but then she grabbed my arm once more.

"So, let me tell you what we're gonna do then. First, well, let's figure out which pairs of colors vibrate for you when juxtaposed against one another. I don't just mean the primary colors, I mean you'll mix ones up that are intrinsically pleasing to *your* eyes. Then we'll use those to construct your own personal color wheel."

"You mean like the chart on the wall there? That maybe the two tones that vibrate against one another for me are opposites . . . like red and green are for everyone else?"

"Possibly . . . if you forget that part about *everyone else*. We've got a *linguistic* problem here, kid. The entire English-speaking population uses terms like *red* and *green* as if we're all seeing the same things, which, color-blind or otherwise, we're probably not capable of doing anyway."

I wasn't exactly following her fully. At age fifteen, I hadn't thought much about the difference between linguistically labeling a perceived color and the color itself. Dorothy Ives had. She was painfully aware of the fact that we all see reality a bit differently, even though in our society we're likely to gloss over those differences and therefore gag our unique voices. Dorothy, twenty years my senior, had lived on the street as a teenager— no family or home or high school degree. Somehow, she tenaciously held onto her dream of becoming an artist and ultimately achieved a Bachelor and Master of Fine Arts during midlife, largely because of the uniqueness of her vision. She felt it to be her moral imperative to open others up to seeing the world in their own unique ways.

Instinctively, she grabbed a lock of her unruly hair, remembering that I had once voiced my amusement over how many times it had been bottle-tinted.

"You see, kid, you've been calling this color *red* ever since you could read *r-e-d* on the label of a Clairol bottle, but whatever you've seen all those years isn't what I see when I say *red*. So just put aside the labels, put them aside, they don't help. Let me put it to you this way: Now that you've tested color-blind, you have even less reason than the rest of the students to put much stock in labels. So let's get on with it."

The rest of the semester, I was released from other assignments to concentrate on an independent scientific study of color under Dorothy's guidance. She brought in piles of books on color from her own home and from the branch campus of the state university. They covered seemingly disparate topics, which we struggled to stitch together: the physics of light; the neurobiology of color perception; the cultural classification of color groupings; the chemical mixing of colors. She would leaf through one of the books until something struck her as relevant; then she'd assign me to attempt some experiment that might reveal our physiological differences in color perception, or at least our divergent aesthetic preferences.

It was the first time I had ever felt the excitement that experimentation and scientific discovery could bring, and it happened in an art class in the basement of a rather ordinary public school in Gary, Indiana. Together, Dorothy Ives and I began to generate new hypotheses, which we tested through our own experimental designs. Whenever we came upon a new lead for our visual experiments, she would start to wildly tap her fingers on the desk and think

21 ☞

aloud, while the students assigned to other tasks would look at us as if we had both gone mad.

On one occasion, a senior—an impeccably dressed girl who was already a professional model—nudged me as the bell rang, suggesting that we talk as we walked together from what she called "Dorothy's madhouse" to our math class on the second floor.

"Gary, I think you should know that some of the students feel you're getting too much attention from Dorothy. She doesn't notice what some of the rest of us are doing. It wouldn't be so bad if we thought you were painting something cool, like something that could eventually go up in the halls to represent our class for posterity. Well, I hate to break you the news, but your paintings are getting even uglier than they were before."

I felt blood rushing into my forehead and cheeks. I just stared at her, unable to speak. I noticed for the first time that her fire engine–red pleated skirt was perfectly matched by a red-and-white blouse, knee socks, and hair ribbons. It struck me that she looked a little like one of the many millions of mass-produced Coke cans.

"What did you say?" I asked, coming out of my daydream, abandoning my thought about whether her kisses tasted like Coca-Cola.

"I said that we don't think Dorothy should spend so much time on helping a student who can't even paint anything pretty."

"You don't get it, do you? When Dorothy proclaims that beauty is relief from monotony, she's telling us to quit worrying about what's *pretty* or *cute*. And she isn't just talking about what's happening inside her classroom.

She's urging us to do something special with our lives. What you call *pretty* is some cookie-cutter version of art; what you call *cute* is boring to me. I'm sorry you feel Dorothy's ignoring you and wasting time on me, but at least I'll never end up boring you by selling bad art to K-Mart!"

Her eyes flashed with anger, or perhaps with hurt. As she slipped away from me into our math class, I suddenly felt terrible for having tried to defend myself by putting her down. Impulsively, I decided to skip class and walked out the back door of the school into the wood-covered dunes to the north. Within a half hour's time, I was on the Lake Michigan shoreline, drawing colorless figures with a stick in the sand, letting the waves wash them all away.

Over that long and lonely semester, Dorothy Ives helped me be less reactive to what others thought of me; she offered me a means of understanding the ecology of colors as they interacted with one another and with my own eyes. Each hue's salience was embedded in a particular context, peculiar to the viewer's perceptual capacities and cultural biases. Sure, I could say the word *red* when Dorothy's tinted head was silhouetted against a whitewashed school wall. But I couldn't always fathom where *red* resided in the wild. That is, I could not *see red* when some friends giddily pointed out a crimson Indian paintbrush on a rock wall. While they were ooohing and aaahing over its bloom amidst all the greenery of a hanging garden, it hid from me in the variegated leaves, stems, sepals, and bracts crowded together in the mottled light below the rock wall. My friends had to hoist me up so that my nose was pressed up against those crimson blooms before I could see red in the same blotches that they saw from afar.

I later learned that the opposite is also true: certain patterns boldly stood out for me when color-normal people paid them no mind. They were camouflaged for most passersby, hidden in a hodge-podge of hues that obscured them from casual view. And yet it took another twenty years before this differential capacity for pattern recognition became an obvious asset in my work, particularly in my plant conservation work.

When I first began to monitor night-blooming cereus cacti along the U.S.-Mexico border, botanists in the regional Office of Endangered Species were not entirely sure that these plants were truly rare; perhaps, one confided in me, they were just too damn hard to see and survey. About that time, I was reviewing a lot of literature on plant and animal mimicry—a topic that had fascinated Darwin, Bates, and Mueller as much as it did me—and I stumbled upon a commentary by the famous cactologist, Lyman Benson. He conjectured that the lead-colored stems of night-blooming cereus in the Sonoran Desert camouflaged themselves amidst the protective branches of ironwood and creosote bush, whose branches have the same diameters and hues as cactus stems.

I was curious to learn if cacti were cryptically camouflaged for some but not all of humankind. Remembering my experiments with Dorothy Ives, I decided one day to survey two nearly identical desert knolls within the range of night-blooming cacti. One knoll would be surveyed by a team of color-blind botanists, and the other by equally experienced but color-normal colleagues.

Each team meandered over roughly the same size

patch of rocky ground, at about the same pace, during a season when the cacti were not in bud, bloom, or fruit. Bluntly put, the plants were not conspicuously flashing their sexual organs at anyone who happened to come along; they were reticent about exposing themselves. By the time we had accomplished two hours of intensive searching, we had results that were incontestable: My team of color-blind cactus surveyors had encountered over five times the number of cacti that our colleagues had found around the other knoll. The two teams then worked together for several more hours and determined that both hills had roughly the same density of cacti. Over the entire season, I alone encountered more than fifteen times as many camouflaged cacti as all of my color-normal colleagues found collectively.

It is possible that my hit rate may have been elevated by my intense interest in the issue of plant mimicry. Still, I remain confident that my color-skewed capacity for seeing through camouflage was a critical factor in my ability to encounter hidden night-bloomers.

Since that day, I have always wondered if those ancient human populations that remained heterogeneous in their color perception had greater chances of survival than their neighbors. Were they better able to spot cryptically colored poisonous snakes? Could they more quickly detect warriors whose faces and bodies were mottled with muds and vegetable dyes as part of a sit-and-wait-then-strike ambush strategy?

Color-blindness is just one more double-edged sword embedded in our genes. While many folks like me are mostly clueless in the face of certain warning signals—traffic lights changing from green to red on a crowded street corner in a strange town—we are not easily duped by other forms of visual deception. That is exactly why some color-normal fighter pilots during World War II relied on their color-blind cronies riding shotgun in the cockpit to spot camouflaged antiaircraft artillery in the forests below them. Color perception—and relief from deception—allowed the copilots to recognize patterns of potential danger, to avoid them, and to fly on toward less perilous places. Like a pilot with a color-blind sidekick, Dorothy had relied upon my peculiar vision to help her steer our way through enemy territory. It was territory where aesthetic quality had been undermined by a myriad of advertisers' manipulations of our innate sensibilities. When we landed and went our separate ways, I was grateful that we had avoided a fate which both of us agreed was worse than death—living in a society where monotony disguised itself as normalcy.

WHY POETRY NEEDS SCIENCE
Decoding the Songs That Can Help Us Heal

> *Bilinguals may have a more flexible approach to the world . . . from a meta-linguistic awareness of arbitrary, nonphysical aspects of words and the effect of context on the meaning of words. Thus bilinguals may find it easier to encode and access knowledge in diverse ways, and have greater tolerance for ambiguity.*
>
> —Todd I. Lubart, *Handbook of Creativity*

A datura flower, if left unpollinated, is a sad spectacle to behold during the days following the decline of its blooming. It has wilted down to a flaccid, twisted, withered rag—a far cry from the silky evening gown that excited hawkmoths only a few hours before.

That is how I feel about some pursuits of the human mind, which, if kept out of reach of cross-cultural and interdisciplinary exchanges, languish like a withered flower, never to bear fruit.

This is the story of a song-poem that was saved from such a fate only by the wedding of poetry and natural science. By cross-pollinating the linguistic, ethnographic, and poetic understanding of the song with insights from field ecology and neurobiology, we can now celebrate the song-poem in all of its dimensions. We can be thrilled by the loveliness of its imagery and astonished by how it embodies an empirical understanding of plants and animals that modern scientists have only recently gained by other means. Had it not been for a fortuitous convergence of interests among several scholars in the 1980s and 1990s, the multiple layers of meanings for a cluster of O'odham songs would have never been recovered. These songs had come from the dreams of an Akimel O'odham shaman who lived more than a century ago in what is now called the Gila River Indian community. But before we hear one of those songs and explore its many dimensions, I must reflect a moment on the historic events which obscured its multiple meanings, and which later brought them to light.

This song-poem was first written down around 1901, translated from the northern Piman dialect of the O'odham language by José Louis Brennan. Also known as José Lewis, this Tohono O'odham folklorist and self-taught linguist let ethnologist Frank Russell use the poem in his comprehensive study of the Pima Indians of Arizona, which was first published in 1908; he later became the first of his people to prepare texts for the Smithsonian Institution on O'odham language, culture, and ritual oratory. When the poem was recorded, it was considered part of a sequence of jimsonweed or thorn-apple songs performed by an old Piman singer, Vishag Voi'i, or Prairie Falcon Flying. With Russell's help in selecting English equivalents, Brennan transcribed the song in the O'odham language as best as he could, then offered a crude literal translation as well a more poetic translation in English, which follows:

Pima Jimsonweed Song

At the time of the White Dawn;
    At the time of the White Dawn,
I arose and went away.
    At Blue Nightfall I went away.

I ate the thornapple leaves
　　And the leaves made me dizzy.
I drank thornapple flowers
　　And the drink made me stagger.

The hunter, Bow-remaining,
　　He overtook and killed me,
Cut and threw my horns away.
　　The hunter, Reed-remaining,
He overtook and killed me,
　　Cut and threw my feet away.

Now the flies become crazy
　　And they drop with flapping wings.
The drunken butterflies sit
　　With opening and shutting wings.

Unfortunately, the first translators of this song glossed the term *ho'okimal* as *butterfly* instead of *night moth*, inadvertently obscuring how the song's imagery moves back and forth between the moths, their horned caterpillars, and the leaves and flowers of thornapple (jimsonweed). That is because the O'odham term *ho'okimal* broadly refers to all lepidoptera, including both nocturnal moths and diurnal butterflies, but by using the latter term, the plant is divorced from its true (nocturnal) pollinator, a hawkmoth. Unfortunately, every version of this poem printed over the following nine decades has retained the word *butterfly*. This biological imprecision has inadvertently obscured certain layers of meaning that are critically important to understanding the song as a whole.

On the other hand, we can be grateful that Brennan and Russell correctly noted that this *kotadopi ñe'e* (datura song) was considered by the Pima to be related to a genre of *pihol ñe'e* (peyote button songs), both about narcotic plants ceremonially used by the Pima and their relatives to the south. It appears that these songs were introduced from more southerly, Uto-Aztecan tribes in Mexico, who chanted them to bring success to their hunting of sacred deer. Both songs, then, dealt with the power of plant hallucinogens to mediate hunters' relationships with animals. These songs were employed in curing rituals during which *mamakai* (shamans, medicine men, or spiritual healers) attempted to remove the causes of vomiting and dizziness from a sick person's body and soul.

In the 1980s, linguistic anthropologist Donald Bahr began to explore these and other O'odham curing songs as part of his interpretation of Piman shamanism and oral traditions. To his credit, Bahr invited our mutual friend, the late Pima elder Joseph Giff, to help him retranslate the jimsonweed song word by word, and to relate it to other butterfly songs from the same genre. This work provided fresh insights into the songs and offered more linguistic and ethnographic precision than did the original translation.

Unfortunately, Bahr's brilliance as a linguistically rigorous translator is not matched by much literary intuition or interest in the natural world, the very subject of many of the songs and speeches that he has attempted to translate. Bahr's translations have become notorious among both Pima speakers and desert biologists for their lack of ecological precision, generated by his confusing butterflies with moths, ponds with nectar pools, trash with flood-washed organic detritus, and peacocks with macaws. Like many scholarly translators of his generation, Bahr failed to achieve a deeper understanding of the poetic traditions he analyzed simply because he lacked familiarity with the flora and fauna that inspired these poems. It appears that Bahr understood so little of the desert ecology well known to O'odham speakers that he could not even ask them appropriate questions about the rich naturalistic imagery in this and other songs. Unfortunately, Joseph Giff, who taught me to understand many interactions between plants and animals from the perspective of a Pima farmer, was never asked to comment on the biological content of the song by Bahr.

From years of venturing into the Sonoran Desert with O'odham speakers who knew its flora and fauna as well as Joseph Giff did, I soon learned that the songs were not at all about butterflies visiting some weed but about the intriguing interactions between the dangerously beautiful daturas and hawkmoths. While adult hawkmoths visit datura flowers at dusk, dawn, and throughout the night, the larvae feed on datura foliage around the clock. Many O'odham are curious about both the adult and the larval forms of *Manduca* hawkmoths, for they visit sacred datura flowers as larvae and the leaves as adult moths. The O'odham consider *kotadopi* to be a plant that can cause their people to go crazy with hallucinations. The late Laura Kerman—an O'odham woman who would regularly visit my home even into her nineties—would not even brush against the datura plants in my driveway nor inhale their perfumes for fear of going crazy. And yet she was fascinated that certain moths and caterpillars visited such a dangerous plant whenever it was available to them.

Curiously, academically trained desert ecologists didn't devote much attention to the pollination ecology of datura for the first century that they explored the Sonoran Desert. Only within the last three decades did ecologists "discover" that, like monarchs and their milkweeds, *Manduca* moths coevolved with sacred daturas. Their caterpillars are the only creatures known to be capable of consuming and detoxifying the powerfully narcotic alkaloids produced by daturas. The hawkmoth larvae called tomato and tobacco hornworms not only ingest the usually toxic datura leaves, but they sequester the plant's atropine and scopolamine in their flesh to deter potential predators, who suffer greatly if they eat the smallest bit of the moths' alkaloid-laden bodies.

Oddly enough, when adult moths ingest alkaloids from datura, it stimulates a pronounced dizziness, if ecologists have correctly interpreted the behaviors they witness after *Manducas* forage in and around datura flowers. It was not until 1983—eight decades after Prairie Falcon Flying's song was published—that pollination ecologists Verne and Karen Grant described in the scientific literature the dizzy, drunken behavior of adult hawkmoths hovering around datura blossoms—perhaps the same behavior that is alluded to in the O'odham song. The Grants presumed that the moths were hallucinating after having imbibed alkaloids present in datura flower nectar.

However, after reading the Grants' report, many ecologists wondered whether the Grants themselves had hallucinated! These scientists were skeptical that the potent secondary chemicals found in datura foliage and fruit could ever be present in the nectar and pollen of the blossoms. The herbivore-deterring chemical compounds found in the vegetative tissues of desert plants are seldom present in their reproductive tissues, since such an expensive chemical arsenal is hardly ever needed to protect their sexual organs. But when I suggested to several well-known pollination ecologists that the O'odham song corroborated the Grants' study, they scoffed at the suggestion. They thought it highly unlikely that any insects visiting datura flowers could absorb significant doses of alkaloids unless they were also foraging on datura foliage.

Then in 1999 and again in the year 2000, these ecologists humbly admitted that new information had forced them to change their views. Two reputable teams of scientists from other parts of the world reported that humans had suffered mental and physical disorders after consuming honey produced by wasps which had visited (and foraged within) datura blossoms. Their observations confirmed that psychotropic alkaloids are indeed taken up by insect visitors in sufficient quantities to affect animal (including human) behavior. It now seems plausible that the composer of this O'odham song had witnessed such aberrant behavior among pollinators visiting datura blossoms, for his observations predated by at least a century those of the scientists who confirmed this chemically mediated relationship.

And so this convergence of indigenous poetic knowledge and ecological scientific knowledge seems to me to be something worth celebrating. It sent me back to the original O'odham transcript of Prairie Falcon Flying's song, and to Joseph Giff's word-by-word translation of it, for I wanted to arrive at a translation that was at once ecologically precise and literary. With some refinements, it builds on the translation that appears in my chapbook of poems, *Creatures of Habitat,* and in my book of essays, *Cultures of Habitat:*

Sacred Datura-Hawkmoth Song

1.
Stopping for a while in the white of dawn,
Stopping for a while in the white of dawn,
Then rising to move through the valley,
Then rising to move through the valley,
Remembering when the green of the evening fell
    away,
When the green of the evening falls away:

Sacred datura leaves, sacred datura leaves,
Eating you, I dizzily staggered, drunkenly crawled,
Sacred datura blossoms, sacred datura blossoms,
Drinking your nectar, I dizzily, drunkenly flew
    away:

2.
As I hovered, he pursued me, his bow looming
    larger,
His arrow overtaking me, shooting right through
    me,
My horns were cut off, and thrown away.
As I was pierced by the arrow, my guts were spilled,
I fell from the air until my fluttering was stilled,
The horns severed from me had fallen away.

3.
They are bugging me now, crazily buzzing,
bugs are swarming, driving me crazy,

I'm diving, swooping, my wings tucked away,
A night moth drunk on nectar,
I'm so drunk on datura nectar,
I shudder and flutter till it all goes away.

Two years after beginning this translation, I stumbled upon a description of the physiological effects and psychiatric consequences of human ingestion of atropine and scopolamine, the chemicals in datura which are considered to be poisons as well as powerful drugs. I realized that to a remarkable extent the behaviors alluded to in the O'odham song-poem parallel the responses of those who have lived to describe datura-induced hallucinations. I am more and more convinced that Prairie Falcon Flying or some unknown O'odham shaman-poet before him composed this song under the spell of datura's wild chemicals.

As I learned from psychiatric studies, and from interviewing intentional and accidental consumers of datura, it is probably not coincidental that a Pima poet felt as though he were flying and staggering. This happens to most folks who have ingested datura, some of whom claim that they witnessed night skies turning an iridescent green. As the drug takes effect, those under its spell become dizzy and nauseous. They soon begin to suffer from a sort of paranoia, aroused sometimes by the unanticipated appearance of even the slightest of creatures. They may fear that they are being attacked by flies, mosquitoes, or miniscule insects, whose buzzing in their ears is both amplified and distorted. At the peak of their hallucinations, some victims of datura poisoning feel that they have been splayed open, burnt, or brought into the flaming world of the dead. The victims

who are fortunate enough to escape both death and per-
manent disability have described their escape from the
dead and delivery into a brief period of ecstasy. There,
they experience the intense pleasure of being immersed
in a kaleidoscopic world before they lose consciousness
and fall into a comalike state that lasts for hours.

It is not surprising, then, that the creature dwelling
in this song-poem—the one who has imbibed datura
nectar and leaves—recalls a sense of flying in the green
light before dawn, of being chased by a force that tears
him apart, of being tortured by buzzing insects, and of
being left to flutter helplessly to the ground before clos-
ing down.

We will never know whether Prairie Falcon Flying
himself ingested datura, or whether another deer singer
set down this poem. We do know from oral tradition
that O'odham spiritual healers have long known how
to prepare datura roots to induce visions, although most
contemporary medicine men consider the plant to be so
dangerous that it is rarely worth using. However, these
healers may still sing datura songs for children or young
adults who have become sick with nausea, dementia,
insomnia, or dizziness. Through complex cultural meth-
ods of diagnosis, the healers determine that the victims
have at one time or another violated the spirit power
of datura flowers, the moths that visit them, or other
powerful beings. These violations may have occurred
even when the victims were toddlers and unaware of
the potential consequences of their actions.

One account tells how young O'odham girls used to
use the silky white blossoms of datura to fashion dresses
for their dolls. Datura flowers could often be found

in the desert washes and hedgerows where O'odham youngsters played, and they made such lovely gowns that they must have been irresistable. And yet when the girls were found playing with datura, their parents scolded them and hastily confiscated the dolls' dresses. While some children might simply become momentarily saddened or frightened by this reprisal, it appears that others were left with a prolonged feeling of deeper emotional distress.

If this stress later generated psychosomatic maladies, O'odham spiritual healers were brought in to detect the cause of the trouble and to relieve it. They did so by singing a set of songs that acknowledged the violation of some ethical and ecological condition essential to their well-being. The songs often began with a staccato cacophony that recreated the sense of emotional turmoil felt at the time of the scolding, followed by a haunting melody overlaid with words forming a metaphorical riddle, one which could not be consciously solved. Oftentimes, the final images of the song and the final melodic theme subliminally suggested that a peaceful solution could be found. At the same time, the singer took images of flowers and moths cut from deerskin and applied them to the victims' bodies, further encouraging their healing by returning them to peaceful engagement with the creatures of the wilderness world.

I have recently learned that curing songs inhabited by moths, datura, peyote, or deer are common to most cultures forming the Uto-Aztecan family of languages, a family that includes not only the O'odham or Pima, but Yaqui and Huichol as well. In other words, the healing power of restoring relationships with these particular plants and animals was recognized so long ago that it is not restricted to the contemporary O'odham. Nor was it peculiar to the Yaqui shamanism that was distorted and popularized by the late Carlos Castaneda.

Yaqui deer singer Felipe Molina, folklorist Larry Evers, and linguists Ofelia Zepeda and Jane Hill have collectively demonstrated how such songs are used to evoke the Wilderness World. This is also known as the Flower World, the place where ancestral spirits still dwell. The spirits there love the movements of brightly colored and iridescent beings: wildflower blossoms, hummingbirds and butterflies, moths and stars, wildfires and rainbows, dawns and sunsets. To remember and give pleasure to their deceased ancestors, the O'odham, Yaqui, Huichol, and Tarahumara all sing of the dance of pollinators amidst brilliant flowers, of spinning, drifting, and shining objects, of deer bedecked with bouquets and ribbons, standing at peace in the Wilderness World.

Since the words for wildness, health, curing, and healing have the same roots in the O'odham language, Pima medicine men may have selectively used datura-induced contacts with wild creatures to promote a deep healing within those who had been traumatized earlier in their lives. Ironically, modern medical practitioners have recently begun to prescribe small doses of alkaloid extracts from datura to their patients. They offer light doses of these drugs to quell the same intensities of nausea and gastrointestinal spasms that can be triggered by overdosing on datura. Psychiatrists, I suspect, could learn much about the effects of potent plant drugs like atropine and scopolamine simply by more deeply reflecting upon oral traditions regarding datura still extant among Native Americans. Just as some pollination ecologists have come to appreciate the knowledge of plant-pollinator interactions encoded in datura song-poems, neurobiologists might be rewarded by exploring their implications as well.

Indeed, I feel fortunate to live in a time when a growing number of scientists are increasingly inclined to consider the work of poets, and vice versa. And yet, I often wonder why they ever fell out of dialogue with one another at all. For complex reasons, many scientists during the latter half of the twentieth century must have believed that they were the only scholars who could legitimately elucidate the world's truths. At the same time, many poets, novelists, and literary critics became disenchanted with the stories of the wild world; that is to say, natural history became marginal to their core interests. A tide of disengagment between the arts and the sciences rose sometime after World War II and began to ebb around Earth Day in 1970, although its undertow still pulls down some postmodern poets and scientists.

My sense is that the rift between those engaged in scientific and literary pursuits reached its widest dimensions between 1940 and 1970. At that time, only a few agile individuals found ways to straddle the two diverging worlds—Edward Ricketts, Loren Eiseley, Vladimir Nabokov, Archie Carr, Rachel Carson, Peter Matthiessen, and Ursula Le Guin, among others. Perhaps this rift can be likened to the kind of differential pressure found along a geological fault line, where part of a landscape has been forced to slide away, deepening the divide between two sides of a canyon.

One explanation for how this rift formed can be found in Michael Brenson's *Visionaries and Outcasts*. Brenson reminds us of the deep sadness, grief, anger, and alienation felt by many poets and artists after American scientists unleashed the atomic bomb on Japanese civilians and soldiers at the end of World War II. During the Cold War, American bullishness led to additional, almost unbridled, investment in science and technology, while support for aesthetic and moral expressions by artists and writers lagged far behind. The political ideologies of the space race fueled further growth of scientific institutions, but nothing so lavish was offered to the arts and humanities to foster creativity in their domains. By 1968, the U.S. government was investing more than $16 billion a year in resources for scientists and engineers, but less than $8 million a year in opportunities for artists, writers, dancers, and musicians. When contrasted with governmental investment in the National Science Foundation, NASA, and the National Institutes of Health, support for the National Endowment for the Arts and the National Endowment for the Humanities appeared to be a belated, token attempt to keep artists, poets, and social scholars from starving.

And then, as Brenson documents, in the early 1960s a counterbalancing force gained some momentum. Advocates for the arts successfully convinced President Kennedy that extravagant investment in the sciences had created a disequilibrium in America that had diminished its citizens' creative and moral status in the world at large. Beginning with Kennedy and continuing for several more presidencies, political leaders entertained the premise that freedom of artistic expression set America apart from other nations, so that the arts could serve as an indicator of how mature and sophisticated the United States had become.

To gain long-overdue support for fellowships that benefited poets and painters, their supporters argued that outstanding artists, like good scientists, deserved the resources to enable them to experiment and to explore the unknown. Compare, for example, the statements in the 1970s of biomedical researcher Albert Szent-Györgyi with arts advocate James Melchert.

A Noble Prize winner, Szent-Györgyi asserted that "research means going out into the unknown with the hope of finding something new to bring home. If you know what you are going to do, or even find there, then it is not research at all . . ."

Not long after Szent-Györgyi's plea for more support for open-ended experimentation among scientists, Melchert explained to Congress that "what a great many artists do is investigate. For that matter, art can be thought of as aesthetic investigation. Where would science be without research? The same question can be said about art."

By the mid-1980s, artists, writers, and critics found that the National Endowment for the Arts had begun to divert its limited resources away from individual artistic experimentation. Some poets and artists found that by engaging with others in public arts projects that explored the perception of natural environments, they could dip into some of the philanthropic sources that had only supported scientists up until that time. In particular, I think of the re-photography projects in the arid West, where arts photographers went to the very same spots where government-supported documentary photographers first ventured to demonstrate how these pioneers had constructed a peculiar image of the West. While the aesthetic attraction to the natural world felt by many poets, photographers, and painters remained as steadily expressed as it had been for decades, a new dimension emerged in their work: deep moral concern over the ever-more-prevalent desecration of lands and waters they held sacred, and the accelerating extinction of species.

Poets and novelists soon offered up eulogies for fallen species and despoiled places. Some rallied together in defense of the earth and began to embrace the many fresh ecological insights about biodiversity that were emerging out of field research. Instead of further distancing themselves from the sciences, certain artists and writers marveled at all that biologists and physicists were learning about nature. They read all they could about fractals and biomimicry, and their art began to reflect a new scientific literacy.

Around the same period, a few literary critics began to take an interest in the writing of scientists, especially those who displayed an occasional metaphorical flair. By the time the Association for Study of Literature and Environment (ASLE) convened its first international conference in 1995, it seemed as though the Great Rift had disappeared.

This too was cause for celebration, as I sensed at the site of ASLE's first massive gathering of its green literati, at Fort Collins, Colorado. I had already been on the Colorado State University campus for several days before the ASLE conference began, attending the Board of Governors' meeting of the Society for Conservation Biology. Fortuitously, I heard that another group of environmentalists—literary ones—would be meeting next door. When I stumbled into an assembly hall filled with 600-odd aficionados of nature writing, I felt as though I had metamorphosed and ascended into heaven. As SueEllen Campbell, John Elder, and Larry Buell offered their welcoming words at the plenary sessions, I could sense that a critical mass of poets, literary scholars, and environmental historians were no longer wary of being in dialogue with natural scientists. I only wished that more of my colleagues from the conservation biology constituency had gone truant that day and come over to ASLE to hear its leaders' encouraging words about the marriage between poetry and science in service to the other-than-human world.

While many scholars of literature, culture, and history acknowledged the ways they had benefited from reading literary naturalists and natural scientists, I sensed that the percentage of biologists, geologists, or hydrologists who found their work enhanced by reading (or writing) poetry and novels was likely to be far lower. I no longer doubted that those who read, loved, and interpreted Native American literature would appreciate the insights into Prairie Falcon Flying's song-poem that were offered by pollination ecology and neurobiology. But to put it bluntly, I still worried whether pollination ecologists or psychiatrists would find the sacred datura song-poem of much value to their own work, or to their own hearts and souls.

If I were to convince any scientists that their own worldview might be enriched by a greater familiarity with poetry, I would have to demonstrate to them that the fostering of creativity is essential to the advancement of science and that without it modern science will wither like an unpollinated flower. And so I began to ask myself, what has my own practice of science tangibly gained from my own forays into creative writing and my frequent reading of leading poets? Are these pastimes mere diversions, which ultimately distract me from fully exercising the rigors of scientific thought? Or might my interest in other languages, and more specifically in language play, somehow serve to prepare me to be more deeply engaged in metaphorical thinking? If so, does some capacity in metaphorical thinking actually help me generate novel hypotheses to test, or freshly interpret, field conditions and experiments in ways I might not otherwise entertain?

As I pondered the possibility that an engagement with poetry might be of value to scientific inquiry, a preposterous question crossed my mind: Has a hawkmoth ever wondered what good it is to a datura blossom?

WHY SCIENCE NEEDS POETRY
Fitting New Metaphors to Nature When Old Clichés Fail

> *To create consists precisely in not making useless combinations and in making those which are useful and which are only a small minority. Invention is discernment, choice. . . . [Creative ideas] are those which reveal to us unsuspected kinship between other facts, long known, but wrongly believed to be strangers to one another.*

> —H. Poincaré, *The Foundations of Science*

Ramona Mattias always seemed larger than life. When I would visit her in her adobe home in the O'odham village of Ali Jek on the U.S-Mexico border, she would bark strong opinions at me, ask astounding questions, and tell me traditional stories with conviction and passion. She was a large lady—big-boned to begin with but even more massive once diabetes set into her body, reducing her capacity for regular exercise.

During one particular time that I dropped by her home, she was in a massively bad mood. One of our mutual friends who lived in a desert village nearby had just been told that his diabetes was so advanced that he would have to spend the rest of his life hooked up to a kidney dialysis machine every other day. Because he was her distant cousin, Ramona feared that she might soon face the same fate herself.

Uncomfortable with addressing the way our friend's disease might affect her, I tried to shift the conversation to another domain, my current topic of interest—the whereabouts of a basket I had once seen with her that used a plant fiber altogether uncommon in O'odham basketry but known in Seri Indian basketry two hundred miles to the south: the limberbush. Perhaps, I suggested, she could help me ask the elders in her community if they remembered any stories about how that plant became part of their basketry traditions.

She looked straight into my eyes, with tears welling up in her own. "How can you so abruptly change the subject like that? You're always asking about our old stories, our traditional plants. *Can't you see that we're dying?* This diabetes is killing my cousin up the road, and it may kill me some day as well. And you, you have this college education and could use your science degrees to help get us out of this rut of diabetes, but you don't even see that we're dying."

"Ramona, I'm sorry, but you know the doctors say there's no cure for diabetes. Once you get it, it's there for the rest of your life."

"I don't give a damn about those doctors who say there's no cure; my parents and grandparents didn't suffer this way. If they saw early signs of sugar diabetes, they used our desert plants to get rid of them. Why don't you forget about that one old basket and figure out which of our desert foods and medicines can help us get out of this mess?"

I stood in silence, a bit humiliated, a bit unsure that I could really do anything to help. After a while, I stuttered out a conciliatory response:

"Well, like what plants do you think I should look at?"

She tried to subdue her frustration and put it behind us.

"Well, for starters, we have two kinds of skinny cactus

out here. We used to dig their roots up to treat diabetes, one called *witch's wand* in our language, the other called something like *it has potatoes underneath*. Why don't you just go out and find them, see if they really are any good for our current condition? They used to grow between here and the dam out behind the house."

I recognized that both plants were of the night-blooming cereus species I had studied with other goals in mind; she had simply forced me to think of them in another way. With her earlier, angrier words still ringing in my ears, I retreated out her back door and walked a pathway toward the dried-up reservoir a half-mile east of her home. Before long I found the night-bloomers she referred to—one with a fleshy, jicamalike root, another with many potatolike tubercles reminiscent of Jerusalem artichokes. Both had narrow, wand-like stems that weighed hardly anything compared to the mass of roots hidden below the surface.

"A good strategy for a desert plant," I mumbled to myself. "Keep your moisture reserves underground and the drought can't reach them." I wondered, though, how they were used for diabetes.

Ramona didn't know many of the details when I returned to her home, and she suggested I talk with others. But when she died in the year following that visit, and I regretted I hadn't made sufficient progress in learning about this diabetes medicine, her words haunted me enough to get me to drive up to the Gila River Indian Reservation to learn about the plants from a Pima elder with the same last name as Ramona—Sylvester Mattias. Amadeo Rea records our interview with Sylvester in their collaboration, *At the Desert's Green Edge: An Ethnobotany of the Gila River Pima.* During the same visit to the Pima village of Komatke, I told our Pima friend Sally Pablo about Ramona's death and about her request that I pay more attention to diabetes.

"She was right, Gary. If this diabetes keeps advancing, there will be none of us left to tell you any stories."

Soon other O'odham friends died of diabetes as well. In my frustration and grief, I began to read studies by the National Institutes of Health (NIH) and the Indian Health Service (IHS). The O'odham-speaking tribes had the highest known rates of diabetes in the world. But despite the technical excellence of the NIH research in documenting how diabetes affected their overall health risks, something was amiss: Over the twenty-five years of intensive study of the "Pima metabolism," the prevalence of diabetes continued to rise, elders continued to die, and the "adult onset" of diabetes continued to kick in at younger and younger ages. More than a quarter-century of research had not stemmed the tide of suffering, even though the NIH and IHS have spent more than $5,000 of research funds per Pima family over that period of time.

"We feel like rats in their laboratory," Sally Pablo confided to me one of the last times I saw her before she, too, died a premature death. "I've followed all their recommendations, but I've lost my sight, had a kidney transplant, dealt with dialysis and drugs to the point that I'm worn down. Gary, can't you help us figure this out like Ramona asked? It's as if our traditional desert foods used to prevent diabetes but now that we're losing those foods from our daily lives, our bodies are so out of balance that we're losing ourselves."

The failure of all this research to positively affect the problem it was supposed to be solving is not peculiar to this issue. As John Horgan describes so well in *The End of Science,* bright, well-equipped scientists are meeting the limits of what they can accomplish *as long as they stay within the paradigms of their own disciplines.* Horgan builds on the critique initiated by eminent biologist Bentley Glass, the former president of the American Association for the Advancement of Science: "We have been so impressed by the undeniable acceleration in the rate of magnificent achievements that we have scarcely noticed that we are well into an era of diminishing returns."

As Horgan amply documents, few prominent scientists today feel that they have been involved in even a single creative breakthrough of the magnitude of those catalyzed by Darwin, Mendel, or Einstein, despite the fact that scientists now have more sophisticated tools at their disposal and more financial support for their work than at any point in human history. Their failure, Horgan argues, has not been due to technique or technology, but to the way they define their problems. In the case of the NIH researchers, they spent hundreds of millions of dollars looking for what they believed was *the* thrifty gene which made the Pima, Papago, and their other Native American kin vulnerable to adult-onset diabetes. When precise genetic markers were finally identified, they helped to determine that there is no single gene conferring susceptibility to this type of diabetes—there are many different genes, which interact with diet and the environment in ways that can trigger the expresson of diabetes among Native Americans. At the same time, anthropologists working among contemporary hunter-gatherers and paleo-nutritionists studying the bones of early humans debunked the theory embraced by the NIH that a single thrifty gene had helped ancestral hunter-gatherers by allowing them to survive famines. This theory maintains that a thrifty gene that would help hunter-gatherers quickly store energy as fat during times of abundance would help them survive later famine. But when agriculture allowed storable carbohydrates to be available year-round, this "adaptation" for hunter-gatherer societies became a "maladaptation" for agricultural and urban populations. Ironically, however, there is virtually no evidence to suggest that famine or starvation affected many people until they began to live in sedentary agricultural communities. In short, the NIH had spent millions of dollars on the basis of faulty genetic, evolutionary, ethnographic, and nutritional premises.

Worse yet, all of their work had fostered a certain fatalism among the Pima and other O'odham. The research had been explained to them in a way that implied no escape from a genetically predetermined fate; that is, it seemed there was nothing they could do to prevent diabetes because it was hardwired into their bodies.

Of course, this view sidesteps the fact that up until 1950 diabetes was hardly noticed in Pima or other O'odham villages because diet and exercise controlled or prevented it. (The NIH has only sponsored three studies of the traditional Pima diet over the last quarter century, and two of them were again based on faulty premises.) I became interested—at Sally Pablo's urging—in trying to figure out exactly *what* in the traditional diet had historically protected the Pima from diabetes.

Luckily, I crossed paths with an Australian nutritional chemist who had done a similar project with the desert foods of Australian aborigines. Jennie Brand-Miller had discovered something that NIH researchers had not even imagined—that many desert foods lower blood sugar levels and increase insulin sensitivity for diabetes-prone indigenous peoples more than they do for Anglos. But as she learned more about diet change among various ethnic groups, she realized that whenever a people who had eaten desert foods for centuries had been displaced or had lost their traditional diet, their incidence of diabetes skyrocketed. This was not only true for Australian aborigines and the Pima, Papago, and other O'odham, but also for Yemenite and Ethiopian Jews who moved to Israel and adopted the Westernized diet of Israeli Jews.

When Jennie visited me at my home in Tucson one autumn, she offered to analyze native foods from the Sonoran Desert for me if I would help her answer a question she had been brooding over. As we stood in my kitchen, sipping prickly pear cactus cocktails, Jennie loosened up enough to wonder aloud, "Gary, you're a desert ecologist, so you might be able to answer this question that has been plaguing me, since my training in nutritional chemistry doesn't suit me for it. Coming here to see another desert, I'm curious to learn what the flora of various deserts have in common that would have allowed the plant foods derived from them to effectively control diabetes. What I'm asking is this: Is there anything that most desert plants need to survive that then might have benefits in the diets of people now prone to diabetes?"

Her question took me so much by surprise that I could offer her no immediate answer, only silence. Not until later did I see Jennie's challenge as a perfect example of what Jacob Bronowski explored in *The Ascent of Man:* "That is the essence of science: ask an impertinent question, and you are on the way to the pertinent answer." Jennie set the stage for another moment of cross-pollination that has profoundly affected my life.

As much as I wanted to, I couldn't get a handle on Jennie's question as long as I approached it piecemeal, for each desert plant had many adaptations to its environment, and each involved different sets of chemicals in its leaves, fruits, roots, and seeds. For weeks I mulled over analyses of the chemicals produced by desert plants as protection against damaging radiation, drought, insects, and large herbivores, but no pattern emerged that gave me any insight into Jennie's question

Then one evening I happened to be reading poems from Amy Clampitt's fine book *What the Light Was Like.* In the section called "The Hinterland" I found a poem about Amy's response to placing her deceased brother's ashes in an urn just as the migration of monarchs was taking place. In "Urn-Burial and the Butterfly Migration," she compares the urn and her brother's life with the sheathlike cocoon from which the butterfly emerges, soon to migrate:

> An urn of breathing jade, its
> gilt-embossed exterior the
> intact foreboding of a future
> intricately contained, jet-
> veined, spangle-margined,
> birth-wet russet of the air-

traveling monarch emerging
from a torpid chrysalis. Oh
we know nothing

of the universe we move through!
My dead brother, when we were
kids, fed milkweed caterpillars
in Mason jars, kept bees, ogled
the cosmos through a backyard
telescope. But then the rigor
of becoming throttled our pure
ignorance to mere haste
toward something else.

What struck me about Clampitt's poem was the way
her simile moved both ways, from brother to butterfly
as well as butterfly to brother. That is, her bi-directional
comparisons allowed us to learn more about both a
torpid larvae in chrysalis and her brother's ashes in the
ornamental urn, rather than one at the expense of the
other. Rather than metaphorically describing a fog that
"comes / on little cat feet" without telling anything more
of the cat, Clampitt's comparisons found pots of gold at
both ends of her verbal rainbows.

That was exactly what I needed to do to solve Jennie's
riddle. Rather than analyzing how chemicals protect a
desert plant from stress and then seeing if any of the
same chemicals produced a means of reducing the dia-
betic stresses on a human's metabolism, I needed to run
the simile in both directions simultaneously until previ-
ously unnoticed patterns became obvious.

As soon as I shifted my perception of the problem
in this manner, a fresh reading of the scientific literature
made part of the solution immediately obvious. Jennie's

work had shown that the foods that best controlled diabetes were digested and absorbed slowly, so that their complex carbohydrates broke down to simple sugars over many hours, without stressing the human body's capacity to produce and use insulin. My botanical colleagues Rob Robichaux and Park Nobel had shown that the tissues of many desert plants survive drought stress by slowing down the release of water to (and through) photosynthetically active cells. With the simile of slowed digestion and slowed water loss in mind, I looked at the studies of one desert plant traditionally eaten, the prickly pear.

Suddenly, two bodies of scientific literature—one on the stress physiology of cactus, the other on human metabolic responses to eating prickly pear pads—came into perfect alignment: The same extracellular polysaccharide mucilages that slowed down water loss from a prickly pear pad were what slowed down digestion and absorption of sugars, helping to control diabetes in humans!

Then I looked at the tiny seeds of the mintlike chia plant in a similar manner, for this was one of the desert grains which had fallen out of the diet of Pima and Tohono O'odham families about the time that diabetes had become prevalent. A similar pattern emerged: The gooey mucilages that copiously exuded from the seedcoats of chia with germination then developed into a gel-like sac around the seedlings, protecting them from desiccation. When mixed with a little water, chia seeds could either become the fleece-like covering on chia pet pottery or they could become an effective food for increasing the endurance and controlling the blood sugars of Indian runners.

As Clampitt's poem inspired me to craft similes to interpret other desert plants as well, a larger pattern emerged. It made evolutionary sense for plants in hot, dry lands to be particularly rich in substances that slowed the loss of water from plants and reduced their susceptibility to dessication and death during drought. That these same substances—polysaccharide mucilages, inulins, galactomannins, and pectins—also slowed the rise of blood sugar levels of human inhabitants of deserts meant that they inadvertently offered protection from diabetes. Although similes were initially conceived as being unidirectional, the metaphorical value flowed in two directions: in this case, one way for plants, another for humans.

I continued to read anything I could get my hands on by Amy Clampitt. But whenever I was ready to jump into some tough scientific problem-solving, I'd also seek out other impertinent metaphor makers: Charles Simic, Alberto Rios, Pablo Neruda, and Robert Bly's motley crew of "leaping poets." As larger and larger patterns became obvious—that desert plants everywhere have chemicals that control diabetes; once they are removed from the diets of indigenous people, the people are susceptible to diabetes—I tested their explanatory power through discussions with a number of fine scientists: Boyd Swinburn, an epidemiologist from New Zealand who took up residence with the NIH Indian Diabetes project in Phoenix for two years; Jennie's Australian colleague, pathologist Steve Colagiuri; plant physiologist Suzanne Morse; and physiological ecologist Carlos Martinez del Río. Thanks to them, I became conversant enough with the technical lexicons used by human physiologists and plant physiologists to grow confident that I was looking at scientific matters much as someone who is fully bilingual would look at the issues of literary translation.

As I look back on scientific articles that I coauthored with colleagues in technical journals ranging from *Nature* to *Ecological Applications* to the *American Journal of Clinical Nutrition* to *Conservation Genetics,* I realize that my role in the team effort was most often that of a translator between two disciplines, which then allowed our team to collectively brainstorm until we arrived at a fresh perspective on an old problem. Only much later did I recognize through reading papers in the the *Handbook of Creativity* that there has always been a strong positive link between bilingualism and scientific creativity. As psychologist Todd Lubart explains in the *Handbook,* "The existence of a bilingual advantage [in the creative process] suggests that language as an integral part of culture may restrict the ways that people can creatively conceive of a problem. . . . [In contrast], bilinguals may have a more flexible approach to the world due to a dual linguistic perspective."

Through the bilingual creative process I had stumbled upon while groping to answer Jennie's riddle, I discovered a way to predict which plants on this planet had the potential of producing foods that effectively control diabetes. I sent plants that met these predictions off for analysis in Jennie's lab, and they tested positive for strong effects on blood sugar and insulin.

Over the following three years, I had the good fortune to be able to discuss the implications of this discovery with hundreds of indigenous people who suffer from diabetes, and to train over a thousand Indian Health Service dieticians, doctors, nurses, and community health representatives in how to use native desert foods to prevent diabetes. Boyd Swinburn demonstrated in a clinical experiment at Phoenix Indian School that a complete diet of these foods was sufficient to control diabetes without the supplemental use of medications or altered exercise regimes. Conversely, he determined that a diet consisting of convenience store foods with the same number of calories and the same fat/protein/carbohydrate ratio was all that was needed to trigger diabetes. Inspired by his clinical confirmation of our hypothesis, I gathered a group of Tohono O'odham, Pima, Hopi, and Southern Paiute tribal members for a two-week native foods demonstration project at the National Center for Fitness. Even though we ate as much native foods as we wished each day, all of us improved our blood sugar levels and lost weight.

More recently, my wife, Laurie, and I collaborated with Tohono O'odham Community Action to organize a 240-mile pilgrimage through the Sonoran Desert hinterlands to bring the fruits of these studies to those who most desperately need them. During the pilgrimage, twenty-four of us from four desert cultures subsisted entirely on native foods for two weeks. By the time we arrived at the Arizona-Sonora Desert Museum near Tucson, more than 400 people had joined us in the Desert Walk for Biodiversity, Heritage and Health. We had shaken fast foods from our lives for a while, walked away our fears, and remembered that the slow, simple cures were at our desert doorstep.

None of this would have happened if Ramona Mattias,

Sally Pablo, and Jennie Brand-Miller had failed to ask me the impertinent questions. Perhaps none of it would have happened if I had not been simultaneously looking at night-bloomers, reading Amy Clampitt, and thinking bilingually. Just as important, though, was my access to wonderfully ingenious scientists who would listen to the metaphors and similes that I threw their way, and leap high to catch them. They are not people who wish to practice a science that has merely ground itself into ruts that it cannot get out of; instead, they choose to be involved in science because they are thrilled by the gorgeous patterns they find in this world. They—like you and me—want to see science serve to better the health of human and other-than-human lives on this planet. The cross-pollination of poetry and science may be one of many ways needed to achieve this end. If playfulness is allowed to reinvigorate our inquiries, I doubt that we will ever witness the end of science; instead we may celebrate its reflowering.

HOW METAPHORS CAN SERVE TO CONSERVE
Staving Off the Extinction of Relationships, Saving the Tree of Life

> *If we wait for science to give us all the answers, it will be too late. . . . What then is the moral response that land conservation must take? I believe it is to re-imagine conservation as the expression and defense of all things worth loving in this world. . . . Conservation is the parable through which [one] hopes for a good and noble conclusion. And the storytellers are everyone who has risked entering into a shared sense of love, a shared sense of evolving and maturing a relationship with the land and all the inhabitants of the land.*

—Peter Forbes, *The Great Remembering*

One summer, I found myself puzzled by the apparent rarity of night-blooming cereus cacti down on the U.S.-Mexico border. And so I went out to see if I could find more of them one sizzling day in July. I guess I was hoping that I could encounter a few that had eluded me during my more frequent sunset and midnight strolls along the border. But the glare of the noonday sun offered no solace to anyone or anything fully exposed to its brilliance. I stopped a moment to consider going back home to sit before the swamp cooler in my office. Instead, I drank the last of the water I had in my water bottle, wondering whether I should have brought along a larger canteen, and continued on through the heat.

I had left our little house in the big desert rather abruptly that morning, opting for fieldwork over writing time. The reason was simple: I found that I was too cramped up to write. If I recall that season correctly, I believe I was on deadline for two journal essays and a book, but I hadn't been able to write an engaging sentence for days. I felt as though my creative juices were ebbing. I turned to field biology to pass the time, and perhaps to consider whether it might turn out to be my only option for satisfying work in the future. Perhaps I should simply concede that my pursuit of creative writing frequently led me into bouts of restlessness and extended anxiety while the practice of field science seemed to anchor me.

Then, as I ambled along through the stinking hot desert, worried more about my own fate than about the fate of a threatened cactus, I began to ask myself big questions about the relative value of scientific and artistic pursuits:

"What difference would it make," I wondered aloud, "if I chose to communicate about the natural world only through the rubric of the technical sciences and gave up on trying to express myself through poetry and art? Who would care? Would my words explain anything less to others about how nature works and why the continued integrity of wildlands matters? Isn't the technical lexicon of the sciences sufficiently precise to describe just about any phenomenon that we wish to record? In fact, aren't many poets sprinkling scientific terms into their writing because of their very precision?"

I took a deep, dry breath, stepped over a rattlesnake carcass that had already rotted down to dry bones, and continued my soliloquy:

"For that matter, isn't the logic of science so airtight

that any rationale for protecting the natural world can best be articulated through applying its objective set of principles to set conservation priorities? Why," I asked, as if resting my case before the judge, or the desert itself, "is there even a need for metaphorical language when science has advanced so far in its precision and expanded so broadly its field of inquiry?"

And that is when I found something on the top of a knoll that I had hiked upon a dozen times before without ever noticing. I stumbled upon a night-blooming cactus in the most improbable place: out in the open—just as I was—and bearing the full brunt of the noonday sun. As I shifted my attention to this two-foot-tall, wandlike cactus stem, I noticed that it looked as though it had been burnt on its south-facing ribs. But not burnt by a wildfire, I quickly surmised, for there were no traces of flames having moved across the desert floor, scorching nearby shrubs as they went. No, the burns it had suffered were much the same as those that I might suffer if I stayed out here all summer long: sunburn.

Sunburn? A sunburnt cactus? A cactus could suffer from sunburn? As I took a step back from the scene to reconsider my initial conclusion, I noticed something else that had escaped my eye when I had reached the top of the knoll. About eight to ten feet away from the cactus lay the remnants of a giant ironwood tree. It had recently been felled by a chainsaw, and its trunk had been sliced into logs that had been hauled away, leaving a brush-pile of branches with wilted leaves in its stead.

I felt as though I were reconstructing the scene of a crime: the now-sunburnt cactus had probably germinated and lived its entire life beneath the shade of that ironwood tree. Harbored beneath such a nurse plant, the cactus had never, until quite recently, experienced anything but diffuse, mottled sunlight. Suddenly, its cover was blown when the chainsaw massacre occurred.

Within a month of this first sighting, the sunburnt cereus died.

As I tried to record pertinent data in my field journal, I felt as though I was doing an inadequate job writing an autopsy. The measurements I made and the context I recorded did not seem to fully describe what had happened. A cactus tagged number 29 had one 22-inch-tall stem which was less than an inch in diameter. It was first discovered on May 15, a quarter-mile south of the U.S.-Mexico border. By the time I first noticed the cactus, its epidermal tissue was flaking off, whitened as if burnt on its southern exposure. By June 15, the plant looked dead, its epidermis entirely shed, its ribs exposed and broken, apparently due to trampling by livestock, given the hoofprints nearby.

When I reread my field notes at home, I felt dissatisfied. This plant had been killed by something larger than the hoof of a horse, by something more penetrating than sunburn. To grasp the ultimate cause of death, I had to turn to a metaphor I had first seen in print in a popular essay by Dan Janzen in *Natural History* magazine: the extinction of relationships. Up until that time, biologists had reserved the term *extinction* for the permanent loss of species; Janzen had metaphorically extended it to describe the cessation of vital ecological interactions.

Once its nurse plant was killed off, that night-blooming cereus could simply not stand to be left all alone. Worse yet, the entire population of night-blooming

cacti scattered along the border had no chance of sur-
viving should woodcutters remove all the other iron-
woods towering over them. Terminate the relationship
between nurse-tree canopy and understory cactus, or
between night-blooming cactus and night-flying moth,
and the cactus population loses its viability for long-
term survival.

At most, damaging radiation and trampling were
the proximate causes of death; the ultimate cause was
best expressed by the poignant metaphor that Dan
Janzen had penned in Costa Rica years before. If I truly
wanted to save that borderline cactus population, I had
to take Janzen's extinction of relationships metaphor as
the larger truth. What's more, I had to be dead serious
about how I used that metaphor to describe what was
happening in the world around me, because it more
precisely recorded just what had occurred on the vol-
canic knoll than all the quantitative measurements and
other data I had so diligently recorded there.

Like the cactus, we stand in relation to others or we
succumb to failure. Curiously, I am propped up by some
relationships that I may have believed I only imagined
at first, but they have been proven to be physically tan-
gible for me and for others. That is to say, as Bill Stafford
once suggested, that we benefit from "stories that could
be true"—that we recognize new possibilities in the world
through our imaginations, and then we see that they
become manifest in other ways.

Physicist Leo Kadanoff once described the imagi-
native element of science in a similar manner: "It is
an experience like no other experience I can describe,
the best thing that can happen to a scientist, realizing

that something that's happening in his or her mind exactly corresponds to something that happens in nature." Once I realized that the extinction of ecological relationships was the cause of decline for many species around me, I could imagine that a focus on these relationships might leverage more effective conservation actions than relying on piecemeal, species-by-species approaches.

In November of 1991, ethnobotanist Mark Plotkin, green businessman Jim Hills, and I brought together a dozen scientists, artists, writers, and businessmen for a brainstorming session on the desert coast of the Sea of Cortez. As the writers listened to what the scientists had to say, they challenged them on the language they used, and in a couple of cases asked that the scientists consider entirely new research questions.

From the outset, the writers and artists asked the scientists to speak from their hearts about the issue. When the scientists stated that "the species *Olneya tesota* represented a monotypic species, and was a narrowly restricted endemic," the nonscientists recoiled. The writers offered another way of speaking about the tree: "Not closely related to any other tree, ironwood grows only in the Sonoran Desert and if lost there it will be lost to the world."

When another scientist called *Olneya* "desert old growth" because radiocarbon dates on wood from old trees suggested ages of 800 years or more, another writer offered a tip that came from having worked for several years in the Pacific Northwest: "Old growth sounds like the stubble on the chin of some graying wino. . . . The concern is that we must learn to respect our elders, even if they are of another species. These ironwoods and organpipes may have sprouted before Europeans arrived on this continent! They're ancient cactus and legume forests, not 'old growth!'"

And when ecologists called ironwood a "habitat-modifying keystone species of xeroriparian corridors," the artists began to draw a desert tree of life while the writers called it "a wellspring of diversity, a safe haven for over five hundred other kinds of desert life, a protective harbor for endangered owls, desert sheep, pronghorn antelope, and fragrant cacti."

"Burn beautiful images into our memories," they pleaded. "Remind us why we should care."

Perhaps the most brilliant brainstorming came just before the meeting broke up. Mark Plotkin, founder of the Amazon Conservation Team, asked us, "All this mesquite and ironwood charcoal that is being mined from the desert—where is it going to?"

Sonoran forester Gilberto Solís spoke up: "That's what those of us in Mexico can't do anything about: the end users. Trainloads of ironwood and mesquite charcoal leave our country for high-end restaurants in Los Angeles and San Francisco. They're beyond our reach."

"Maybe not!" Mark exclaimed. "If the more environmentally conscious chefs at those restaurants are contacted to publicly proclaim that they won't use any more mesquite charcoal until slower-growing ironwood is kept out of it, thousands of food devotees will follow suit."

I was stunned; we had never thought of contacting chefs before, but they were key to changing consumer trends. The dialogue between Mark and Gilberto had laid out the path before us. The brainstorming that followed was so exhilarating that I felt as though I was getting some contact high simply from being around these creative thinkers. Within a month's time, environmental journalist Jane Holtz Kay had an ironwood story on the front page of the *San Francisco Examiner,* with a photo of Alice Waters, the reknowned owner-chef of *Chez Panisse* expressing her dismay that chefs had no idea that the bags they purchased labeled "100 percent mesquite charcoal" were causing such destruction in ironwood forests. Once such a food trendsetter as *Chez Panisse* cancelled its order for mesquite charcoal, other restaurants rapidly followed. Within ten days, *Newsweek* ran a short feature on the problem, and forty other newspapers in the border states ran their own articles.

Within a matter of months, an executive from a charcoal-importing company called me to negotiate changes in where and how he sourced his materials, for his sales to restaurants in the Bay Area had declined by more than 15 percent since Jane's article ran in the *Examiner.*

One evening stands out among all others in convincing me that conservation efforts can indeed benefit from embracing and elaborating metaphors such as the extinction of relationships. If I was ever grateful that I took that metaphor seriously, it was on that evening, eight years after our initial brainstorming session, when I assembled poets and planners, singers and scientists, activists and artists to imagine what we might gain if we made a safer place in the world for cactus, ironwood, hawkmoth, and jimsonweed to persist in relation with one another. It was during the month of May in the year 2000, when night-blooming cereus cacti were breaking bud beneath ironwood trees on the outskirts of Tucson. The poets and musicians who gathered inside an auditorium were encouraging hundreds of folks to heed their call to protect the web of interactions woven into the ancient cactus and legume forests of the Sonoran Desert. Transforming her recent field notes from a visit to the forests around Ragged Top into the prose poem "Under the Influence of Ironwoods," Alison Deming took us into the midst of the place we hoped to protect:

> The road reddens—iron bleeding down from the mountains. The saguaros grow dense—standing sentinel for the unpeopled place. Flicker, thrasher, gnatcatcher, elf owl, cactus wren, inca dove, king-bird, titmouse, jackrabbit, bobcat, packrat, pocket gopher, pocket mouse, peccary, kit fox, coyote, leaf-nosed bat, free-tail bat, sharp-shinned hawk, zone-tailed hawk, whiptail, skink, spadefoot, chuckwalla, vine snake, gopher snake, coachwhip, sidewinder, diamondback, turkey vulture, mule deer, swift.

> And the ironwoods, so quiet, if we had not read their primer we would not see the blue-green haze of their nurture, each one a tangled mess of life—trellising vines, nesting doves, roosting hawks, climbing squirrels, the rain of seeds falling and nestling into leaf thickened ground, nursery of wildflowers and cacti under its shade, burrow of desert tortoise among its roots, profusion of micro-lives in the soil.

Then poet Richard Shelton read his classic, "Requiem for Sonora," about the psychic and emotional costs of seeing more desert disappear within our midst:

what will become of the coyote
with eyes of topaz
moving silently to his undoing
the ocotillo
flagellant of the wind
the deer climbing with dignity
further into the mountains
the huge and delicate saguaro

what will become of those who cannot learn
the terrible knowledge of cities

These poets handled biological facts well in their writing, but they did more than that: They offered an altogether different way of making sense of the world, one perhaps complementary to that of the conservation ecologist. And so I felt encouraged to read from an essay of images gathered from my field notebooks, "Cryptic Cacti on the Borderline," which had emerged as my first attempt at creative writing after finding that sunburnt cactus out in the open:

"In late August—not too long after the rain and its revelation—I stumbled onto a cereus population just after dusk. At first, from a distance, I thought someone had left some flashlights on, dropped out among the desert scub. As I walked closer to some ironwoods and creosote, the flashlights beneath them turned into flowers."

Musicians Paul Mirocha, Mark Holdaway, and Laurie Monti played interludes that gave us time to absorb all the verbal images. Activists such as Kevin Dahl—soon to be put in charge of all Natural Resources for Pima County Parks—filled us with jokes and hilarious political commentaries to raise our spirits. I could see in the faces of their listeners fresh recognition of what it was that we belonged to.

On the way home from that event—Writers Rally Around the Ironwood Tree—I stopped my car along the road where I knew that a cactus was close to blooming in the shadows of a stately old ironwood tree. There was but one delicate flower evident that night, but it was fragrant with hope itself.

The day after the rally, I received a call from Maeveen Behan, a tireless lawyer who guided the ambitious Sonoran Desert Conservation Plan on behalf of the Pima County Board of Commissioners.

"For those of us who are down in the trenches most of the time," she confided to me, "it is such a gift to be inspired as we were last night. We get so drained while trying to build collaborations to get conservation done, that it is essential for us to get recharged now and then."

Those in the trenches included environmental lawyers, endangered species biologists, neighborhood coalition organizers, nature center directors, and those who actually dig deep trenches to rescue and salvage ironwood trees before bulldozers come to blade away their habitat. They do extremely fine work and draw upon the best science they can find to guide them. At they same time, they know that science, in and of itself, is seldom enough to reshape public opinion. People have to feel some visceral connection to an issue to act upon it; they have to have images of an ironwood forest in their heads and hearts.

I was reminded of a passage from Ed Abbey's *Cactus Country* written two decades before the Ol' Desert Rat himself was laid into the ground not far from the shade of an old ironwood. His commentary could have been written from the view of a night-bloomer, it was so close to how one might feel living beneath an ironwood tree:

> It was most excellent to lie there on the cool blue-gray basalt under the tough old tree— ironwoods may be almost as long-lived as Methuselah—and watch the bees buzzing and sipping at the water's edge. The heat and glare beyond our little sheltering bower was terrific, stunning, exhausting; the heat waves looked dense enough to float a boat on. But here in the shade we knew peace of a sort, a happy bliss, ease of limb and mind.

The possibility of experiencing that bliss in the future was guaranteed less than a month after writers rallied around the ironwood tree, when President Bill Clinton designated 129,000 acres of Sonoran Desert as the Ironwood Forest National Monument on June 9, 2000. Secretary Bruce Babbitt had come out to see the ironwood forest and its supporters that spring, and for the first time in history the boards of supervisors for both Pima and Pinal Counties in Arizona endorsed a conservation designation of real magnitude.

Did the voices of poets lending their support to the issue make a discernible difference in this effort? It would be presumptuous to claim that the rally tipped the scales toward public support of the President's proclamation, for it did not. Nevertheless, the rally was the culmination of eight years of collaboration among writers and scientists, artists and activists—collaborations which accomplished far more than scientists could have accomplished alone merely by the sheer scholarly elegance of their arguments based on conservation biology research.

Until writers and artists joined their voices together with other constituencies, ironwoods and the cacti in their shadows had a thorny image problem to overcome. Bluntly put, ironwoods had long been considered stickery, homely, and virtually useless by most of the desert-dwelling public. As I tried to do in the pages of *Orion* magazine in 1991, writers and artists needed to help others see these plants not as ugly ducklings but as swans elegantly adapted to the sea of desert surrounding us.

Only after this perceptual shift occurred did the ironwood and the night-bloomers under its skirts have a chance of being protected. Before this shift, the prospects were dismal: Tens of thousands of ironwoods were being downed by bulldozers everywhere while millions of acres of exotic buffelgrass and thousands of retirement homes were being planted in their stead. As much as 30,000 tons of charcoal made from mesquite and ironwood trunks were being exported from Mexico for use in U.S. restaurants, extracted from the very regions of the desert where the trees naturally grew slowly and suffered low rates of regeneration. The decimation of ironwood forests was so rapid and pervasive that Seri Indian communities which had gathered dead ironwood fragments off the desert floor for centuries could no longer find any wood to salvage when they traveled beyond their reservations. In short, their neighbors had cut down one out of every six living desert legume trees and taken away most of the dead wood as well. Any large tree within reach of woodcutters was vanishing

71 ✦

from sight, and once their protective canopies were obliterated, dozens of species as vulnerable as night-bloomers declined as well.

When I began to fathom the magnitude of destruction of ironwood habitat on both sides of the border, it was difficult for me to imagine what could be done about it. Nevertheless, I did have enough faith to know that if I brought together a creative cohort of scientists and writers, a fresh vision of what could be done might emerge.

Over the next few years, a dozen talented writers from both sides of the border lent their hands to this effort, leaving readers with a new sense of awe regarding the dynamics among ironwoods, cacti, insects, and birds. I became particularly excited when Seri Indian woodcarver Humberto Roberto Morales wrote a bilingual poem about the ironwood's importance to his people and distributed it as a broadside in the capital city of Sonora. In the poem, he reminded others that as "a people who are themselves on the edge of survival / we know that it is this [tree] that gives us life." In addition, one of Mexico's most esteemed biologists, Albert Búrquez Montijo, tried his hand at the genre of literary natural history in an essay called "Arbol de la Vida" (Tree of Life), which was published in several Mexican newspapers and translated into English for use in *The Ironwood Primer:*

> Ironwood, in the secret garden of the desert, searches the depths of the soil for precious fertility and subterranean pockets of moisture. Ironically, these treasures are what the tree itself distills and offers as a gift to other plants by virtue of its own

activities: over the years, its innumerable leaves have decomposed and transformed themselves into a thick spongy layer of water-holding organic matter, rich in nutrients otherwise rare on the desert floor.

Soon a poster was designed by artist Paul Mirocha to reinforce this *arbol de vida* concept, using imagery from Mexican folk art. It was seen in craft shops, tourist centers, and environmental protection offices across Sonora, Arizona, and Baja California, prompting one Mexican government official to call it the single most effective graphic ever to be used in the service of conservation in northern Mexico. Exequiel Escurra, Mexico's environmental minister, came up with a new conservation category for ironwood and other culturally significant plants, granting them special protection status nationally because of their roles as keystone species. Mexico also granted the Seri Indians exclusive rights to their traditional ironwood carving designs through a collective trademark, with some officials acknowledging that the Seri ironwood harvest was one of the few harvests of this species that appeared to be sustainable.

If beautiful words and graphic images had not helped a small cadre of environmental scientists and cultural rights activists to reach larger audiences, ironwood would still be put under the plow and the chainsaw blade and into the oven as it had been in the past. Instead, this tree of life and those who cluster around it now have a collective identity, one that was hardly articulated until poets and artists added their talents to those of the scientists.

In 2002, the American public told pollsters that the

most pressing environmental issue affecting this planet was no longer rainforest destruction, pollution, or over-population, it is the loss of places dear to people's hearts. Ironwood forests have become one of those places of heartfelt concern.

I do not believe that it is mere coincidence that the loss of places has been a major recurring theme of the writings of nature poets, novelists, and essayists over the last quarter-century. From W. S. Merwin and Homero Aridjis to Naomi Shihab Nye, Gustaf Sobin, and Cormac McCarthy, our best contemporary writers have helped us grieve the loss of our places of the heart so that we might have the gumption to prevent further losses of this nature. Their seamless stitching of biological themes in their literary masterworks has made these losses of place all the more palpable and, ultimately, intolerable. In the case of the night-blooming cereus and its habitat amidst the venerable ironwoods, marrying science with poetry has generated enough human compassion that these special places may somehow be kept safe for the dance of moths among the flowers for decades to come. If such cross-pollination continues to occur, perhaps both the field sciences and the literary arts will continue to bear fruit.

GARY PAUL NABHAN
A PORTRAIT

by Scott Slovic

Canadian anthropologist Terre Satterfield and I arrived
at the Arizona-Sonora Desert Museum late on a warm
April afternoon in 1998. We had come to Tucson to
work on a project supported by the National Science
Foundation and would spend two days interviewing
five distinguished nature writers, asking for their in-
sights into the connections between narrative language
and environmental values. We conducted most of
these meetings at the guest cottage of the University of
Arizona Poetry Center, but Gary Nabhan encouraged us
to meet him out on his home turf, the desert museum
located fifteen minutes west of the city and then at his
nearby house.

Eventually, Terre and I tracked Gary down not in his
dark, crowded office but out in one of the aviaries for
Sonoran Desert birds, where he was conducting an ex-
periment to see if chili powder would work as a natural
pesticide, discouraging crop-eating animals while not
harming native birds. We easily recognized Gary's mop
of dark hair in one of the distant cages. He completed
his tasks, then joined us on the walkway outside the avi-
ary and enthusiastically explained the goal of the study.

Half an hour later we were driving through the desert, following Gary to his rural home. At first glance there was nothing particularly unusual about his stucco house, but then he began to show us the garden in his backyard, the assortment of native plants growing there. This was not an ornamental garden, created for show. Instead, the garden functioned as an informal laboratory, producing chiles and seeds and cactus buds for the scientist's own kitchen and for his intellectual endeavors. Dinner that evening was a stew of assorted, locally grown beans—delicious and unusual. Shortly before the meal began, Gary asked one of his guests to go out in the back yard and "prepare the cholla buds." This meant sweeping a bucketful of green, grape-sized cactus buds across a screen until the spines had been rubbed off. The purpose of this exercise became clear when it was time for dessert: vanilla ice cream topped with chewy chunks of cactus. This was a meal like no other I had ever experienced.

After dinner, Terre and I retreated with Gary to his study, where we pulled out our tape recorders and began discussing what we refer to as "narrative values." Nabhan explained that he tends to be relatively unconscious of the values aspect of his writing when he works in the storytelling mode, but that he clarifies his own thinking by exploring topics through story. He told us, "For instance, in *Cultures of Habitat* I knew there were some inherent parallels between cultural diversity and biological diversity. So I basically said, 'I'm going to try to immerse myself in the tensions between those two and see what new insights emerge.' So I don't necessarily predict where an essay is going to go, or for that matter where a book is going to go in advance. I don't outline it in advance, but I do put together pieces to see what juxtapositions are fruitful, and then I pursue those." For more than two hours, we discussed story as a special "zone of tension," as a form of language that enables the illumination of scientific information and the articulation of personal and cultural values. Occasionally other dinner guests wandered into the office and joined the conversation, sometimes introducing temporary tangents. The interview transcript, before editing, reads like Samuel Beckett's transcription of James Joyce's *Finnegans Wake*, complete with the author's vagrant thoughts indicative of his attention deficit disorder.

The revised version of our conversation appears in the book *What's Nature Worth? Narrative Expressions of Environmental Values*, but the full context of our 1998 meeting—complete with chile experiments at the desert museum, cholla buds on ice cream, and an interview setting that accommodated regular interruptions—did not make it into that formal study. I have the feeling that such situations are regular parts of Gary Nabhan's life, a busy choreography of science, story, cuisine, and friends, all rooted meaningfully and delightfully in place.

Cross-Pollinations: The Marriage of Science and Poetry. Who could write a book like this other than Gary Nabhan? One could count on a single hand—perhaps on two—the scholars and writers capable of making this kind of argument about the intersections between art and literature on one side and empirical science on the other. Nabhan belongs to a rare group of contemporary scientist-writers that includes Edward O. Wilson,

Robert Michael Pyle, Chet Raymo, Bernd Heinrich, Jared Diamond, and Tim Flannery. What's unique about all of these thinkers is their ability to gather and process information as field scientists and to express their findings with the imagistic, metaphorical, and narrative flair of poets, novelists, and literary essayists.

In this book, the author explains the circumstances of several of his important—even breakthrough—discoveries about the processes of his own intellect and about the perennial focus of his research and writing: desert ecology (particularly preservation of native seeds and plants and the pollination processes) and ethnobiology (a blending of ecology and anthropology, focusing on the intersections between human culture and local plants and animals). During the past two decades, Nabhan has authored or coauthored more than fifteen book-length projects and more than 120 articles, resulting in numerous original contributions to research on the Sonoran Desert and the interactions between native people (especially Tohono O'odham, Pima, and Comcáac, or Seri people) and settler cultures in the region. His technical articles have appeared in such publications as *Conservation Biology, Etnoecológica, Journal of Vegetation Science, Ecology of Food and Nutrition, Diversity, Ecological Applications, Nature,* and *Conservation in Practice.* In addition to his extraordinary prolificness as a scientific researcher and author, Nabhan has become one of the leading figures in the field of environmental literature, with works such as *The Desert Smells Like Rain* (1982), *Gathering the Desert* (1985), *Songbirds, Truffles, and Wolves* (1993), *The Geography of Childhood* (1994, with Stephen Trimble), *Cultures of Habitat* (1997), and *Coming Home to*

Eat (2002) becoming standard works in bibliographies of important nature writing and frequently taught in college writing and literature courses.

But in all of his work, despite the eloquent autobiographical passages in many of the literary books, there have been few moments of overarching self-analysis, explanations of his habits of mind and tendencies of expression. This *Credo* thus offers a rare, explicit glimpse into the strategies and styles of one of the most unusual and productive contemporary American environmental scientists and writers. He lists his primary tendency in "An Overture" in this book: his mothlike sense of freedom in moving from idea to idea, across disciplines and cultures, which matches the moth's flight from plant to plant. Just as certain moths travel among different kinds of plants, Nabhan freely alternates between field science and literary art, recognizing no intrinsic boundary between these disciplines; he feels a deep fascination with both human-altered landscapes and wild, nonhuman places; he savors both solitude and playful, rigorous interaction with other people (laypeople, scientists, resource managers, and artists); and he takes particular delight in the cross-cultural interactions afforded by reading and by his conversational ability in several Native American and European languages. Each of these categories of intellectual movement and interaction is an example of what Nabhan calls, metaphorically, cross-pollination.

Those who know Nabhan's work will find this a truly apt and meaningful metaphor. In 1996, Nabhan and entomologist Stephen L. Buchmann collaborated on a book called *The Forgotten Pollinators*, in which they argued that pollination is "one of the world's most vital processes linking plants and animals—a process that not only keeps us fed and clothed but feeds our domesticated animals and their wild cousins as well. Even more important, it keeps the verdant world, that delicate film of life around us known as the biosphere, running with endless cycles, feedback loops, and checks and balances." In the most literal sense, then, pollination functions as a crucial, life-supporting process. The biological process of pollination frequently requires cross-species interactions between certain kinds of animals and plants. The particular type of pollination that Nabhan takes as a metaphor of several of his own intellectual tendencies is this cross-species interlinking of unlikely and seemingly incompatible forms of thought and expression. Although Nabhan's own cross-pollinating skills are clearly beyond the ordinary, his self-analysis functions as an exhortation to readers and fellow writers. Do not remain casually in your place of comfort! Wander afield, like a moth or a hummingbird, or like a vagrant nature writer! Savor the richness that results from interaction!

A quintessential example of Nabhan's cross-disciplinary, ethnobotanical nature writing—at once scientific and literary—is the essay "Hornworm's Homeground: Conserving the Diversity of Interactions," which appeared in the Spring 1997 issue of *Wings: Essays on Invertebrate Conservation* (a publication of the Xerces Society) and was later collected and expanded, with a slightly altered title ("Hornworm's Home Ground: Conserving Interactions"), in *Cultures of Habitat*. The essay begins with an extended story. As the narrative explains, the author has been working with a crew of observers to monitor the pollination of particular desert flowers in order to assess the possibility that the fruit will ripen in a few months—fruit that will feed people and animals. They have seen few hawkmoths and assume therefore that some of the rarest plants of the Arizona desert will produce few fruit that summer.

While driving one evening to visit some O'odham friends and divert his attention from the "fate of those flowers," the author notices an extraordinary procession of pin-striped worms on the road. He fills a cardboard box with them and shows them to his friends Margie and Remedio Cruz when he arrives at their house. Margie and Remedio explain how they had used such worms as snackfood over the years, roasting them in lard "so that they tasted like popcorn." The septuagenarian Remedio has lived in the southern Arizona desert for his entire life and has witnessed many subtle processes of plant and animal behavior. The university-educated author defers to Remedio's knowledge and asks him to explain where these hornworms have suddenly come from. Remedio predicts that the pollinating moths will soon return to the desert. Nabhan writes: "When and where that might happen, I realized, was something that grade-school-educated Remedio Cruz could predict better than I could. Even with a Ph.D. in desert ecology, I was a novice in observing the life histories of local moths—species which Remedio and Margie had lived among all of their lives." This statement is more than a gesture of humility. The idea that local people who have lived in particular places for many years might have important knowledge about the ecological processes of those places is essential to the field of ethnobiology, and yet many doctors of philosophy may be reluctant to grant authority to ordinary folks. One of the important narrative charms and epistemological arguments of Nabhan's works is the attribution of authority to local ecological knowledge, especially that of native people.

The expanded version of the hornworm essay, in *Cultures of Habitat,* covers some of the same ground as chapter three of this book, showing how, as Nabhan puts it in this volume, "By cross-pollinating the linguistic, ethnographic, and poetic understanding of the song with insights from field ecology and neurobiology, we can now celebrate the song-poem in all of its dimensions." The song in question is an ancient datura hunting song recorded by Pima Indian José Louis Brennan at the beginning of the twentieth century. Thanks to his knowledge of desert pollination processes, developed through scientific observation and conversations with local people, Nabhan "could tell from reading the Piman text that a clumsy translation had obscured a rich ecological message" in the song, so he prepared a new translation himself.

In the fourth and final section of the full essay from *Cultures of Habitat,* Nabhan moves beyond the specific narrative of his hawkmoth field research, his visit to Margie and Remedio Cruz, and his retranslation of the ancient Piman song. He broadens the scope of the discussion to address an issue of fundamental concern to ecologists and conservationists. He argues that indigenous desert dwellers "know about interactions between animals and the major food plants that fill their larders, even when these interactions might seem obscure to the outsider." The problem is that "only now are we beginning to see the conservation value of such local knowledge," and because some native animals and plants have become scarce, fewer and fewer native people have been able to witness the fundamental ecological interactions that would be crucial to knowledge that could support today's conservation efforts. Nabhan's concerns about ecological conservation and cultural preservation, it becomes clear, are thoroughly intertwined. He expresses his fear that "linguistically encoded knowledge about ecological relationships [is] disappearing" and his doubts about the "survival of indigenous knowledge of plant-animal interactions." The essay concludes on a prophetic, admonishing note, quoting ecologist Daniel Janzen, who stated that "what escapes the eye is the most insidious kind of extinction—the extinction of interactions." In a sense, pollination is all about interactions. For Nabhan, ethnobiology, ecology, linguistics, poetry, and storytelling—the many intellectual processes to which he devotes his life—are also dependent on "conserving interactions."

Today, at a time when people in wealthy, industrialized societies travel routinely to distant parts of the world and at home consume products that use natural resources from exotic locales and that are made in far-flung places, we have become increasingly indifferent not only to our immediate environments but to the natural and human resources that go into making our lives possible. Environmental writers are now actively expressing concerns about this tendency toward globalization. One eloquent version of this concern is expressed in Scott Russell Sanders's 1995 essay, "Beneath the Smooth Skin of America." In his essay, Sanders is particularly concerned with the homogenization, the smoothing over, of the local particularities of American communities and landscapes. When we imagine ourselves to be living in a globalized economic and social reality (as many Americans seem to do), it is easy to lose contact with the intellectually refreshing and ecologically meaningful details of particular places in the world, yearning instead for the comfort and predictability of sameness.

Sanders writes, "The earth needs fewer tourists and more inhabitants, it seems to me—fewer people who float about in bubbles of money and more people committed to knowing and tending their home ground." The questions are: What would it mean truly to *inhabit* the places where we dwell? What kind of local knowledge would it take to be fully at home in a particular place on the planet? And might it even be possible, given the attractions and demands of contemporary society, to remove oneself, even to a small extent, from the global economy?

Gary Nabhan is an example of a writer who has spent his career trying to understand how cultures and individuals deeply engage themselves with specific places, learning the natural processes of these places and fitting culture to locality. In the late 1990s, Nabhan conducted a personal experiment to see if he could extricate his household from the global food economy. As an ethnobiologist, he had spent many years studying Sonoran Desert cultures, particularly the Tohono O'odham people, to see how their knowledge of local natural history meshed with their cosmology and language and daily lifestyle. One of the constant emphases in Nabhan's ethnobiological studies and in his literary nature writing is the examination of food and drink. For a period of fifteen months, he tried to rid his household of all food products not grown or raised within 250 miles of his home in Tucson, Arizona. He dedicated himself to growing many of his own foods during the experiment, and what he didn't grow himself, he purchased whenever possible from local farmers. He reports on this experiment in the 2002 book, *Coming Home to Eat: The Pleasures and Politics of Local Foods.*

Here he describes the process of ridding his kitchen of items produced by multinational corporations, showing that he too had become dependent upon the global economy, much like his anticipated readers:

> As I completed my kitchen-cupboard food inventory, I discovered that I had unknowingly purchased foods from six of the world's top ten food and beverage companies: Nestlé, Philip Morris, ConAgra, Pepsi-Co, Coca-Cola, and Mars. . . . Trademarked products from these multinational corporations now generate about one-tenth of all retail sales of foods and beverages across the face of the earth. . . .
>
> However eclectic I thought my tastes in food had become, this inventory revealed that the bulk of my diet had been brought to me by just a handful of food processors and distributors. The contents of my cupboard, fridge, and pantry in Arizona were dominated by agricultural commodities that had been bred, produced, processed, and distributed by the same companies, active

everywhere from Argentina to Zaire. And these companies—while dominating more and more countries' commerce—were becoming fewer in number, as mergers and acquisitions in the food industry recently topped a trillion dollars a year.

It is clear from the beginning of Nabhan's narrative that this experiment is more than a Thoreauvian effort to raise personal consciousness by fully engaging with the nearby and the particular. Nabhan's experiment is rooted in fundamental uneasiness with the social and ecological implications of globalization—uneasiness with the idea of economic and political power becoming centralized in a handful of multinational companies and uneasiness with the social and ecological havoc caused by consumers losing all sense of responsibility for the processes of producing food (and other consumer goods). There is a strong political message in Nabhan's book, an argument in support of radical social reform. Here, for instance, is a passage where he describes the extraordinarily unsustainable use of energy in producing many of the world's food products today:

> The caloric cost of eating was not merely the number of calories produced by the fruit, nuts, meats, and roots we ate; it was also the effort expended in hunting and gathering, in processing and butchering, and most obviously, those we feel as heat while roasting, baking, grilling, or boiling our fare. Archaeologists have estimated that most hunter-gatherers directly consumed a total of some 2,500 to 3,500 calories on the "average" day—including the energy expended while carrying food to cookfires and gathering the wood used in grilling meats and tubers. Of course, averages didn't mean

much to those who foraged in highly seasonal environments. And yet, when the Biodiversity Project newsletter recently published a discussion on my friend Peter Vitousek's estimate that most contemporary Americans require 46,000 calories each day to produce the food they eat, ecologist Stuart Pimm saw it and disagreed.

"Way low," he argued; Vitousek's estimate did not at all cover the many ways in which we consume fossil fuels to transport our groceries, supply our gas stoves, power our food processors and coffeemakers. What Peter and Stuart did agree on is this sobering fact: More than 40 percent of the earth's annual productivity is funneled into feeding just one species, our species, undoubtedly at the expense of the myriad other creatures trying to feed themselves on this wayward ark careening through space.

Nabhan's personal experiment in eating locally is both a critique of the global food economy and a celebration of local community, both the community of human friends with whom he grows crops and prepares meals and from whom he purchases other locally produced foodstuffs and also the local community of plants and animals that have not been bulldozed or chased away by urban sprawl. His book is full of stories of the surprising availability of locally produced foods that made it plausible and even pleasurable for him to remove himself from the global grid. The stories emphasize the pleasures of knowing what edible plants and animals live where you do, and at the same time they suggest that it is possible for ordinary people to avoid consuming genetically engineered foods and foods prepared in ways that exploit land and people in distant parts of the world.

Like most of Nabhan's writings, particularly his *literary* works, *Coming Home to Eat* is gracefully multidimensional, wedding science with experience, story with ethical pronouncement. When journalist Gretel H. Schueller visited Nabhan in Arizona to write about his experiment for an article called "Eat Locally (Think Globally)," she experienced both the pleasures and the politics of this culinary and agricultural study. She notes that "we've grown blind to the bounty of our own backyards. To prove a point, Nabhan stoops over another cactus and says, 'This little one here is the one we get the flower buds off of. They taste like asparagus tips.'" Imagine the pleasure of walking through the countryside, even a landscape as seemingly spare and harsh as the Sonoran Desert, and seeing it as a veritable grocery store and pharmacy! Yet Schueller concludes her article by quoting Nabhan on the ethical aspect of this work: "'Each time we put something in our mouths . . . it's a moral act, whether we admit it or not.'"

Other scholars have previously offered biographies of Gary Nabhan. Sara L. St. Antoine published a detailed portrait in John Elder's 1996 reference work, *American Nature Writers*. Anne Becher provides an entry on Nabhan in *American Environmental Leaders: From Colonial Times to the Present* (2000). Gary himself has offered sketches of his childhood and experiences as a parent in such books as *The Geography of Childhood* and *Cultures of Habitat,* revelations of his fascination with Saint Francis of Assisi in *Songbirds, Truffles, and Wolves* and *Desert Legends,* and a discussion of his Lebanese heritage in *Coming Home to*

Eat. I will therefore give a relatively abbreviated summary of his life story.

Gary Paul Nabhan was born in Gary, Indiana, on March 17, 1952, one of three children of Theodore and Jerri Nabhan. Despite being raised in the Indiana Dunes, on the edge of an industrial Midwestern town just south of Chicago, Nabhan is a scientist and literary artist whose work is associated through and through with the desert Southwest. Already as a teenager, he felt that proper learning should happen out of doors; during high school his truant meanderings among the dunes on the shore of Lake Michigan led to his dropping out of school. After briefly attending Cornell College, Gary made his way to Prescott College, in northern Arizona, where in 1974 he completed a B.A. with an emphasis in environmental biology and Western American literature. During the next decade, he picked up an M.S. in plant science (1978) and a Ph.D. in arid lands resources (1983) at the University of Arizona, developing a scientific specialty in arid lands ethnobotany and agricultural ecology among the Tohono O'odham people of southern Arizona. His scientific work for the past quarter-century has involved a combination of agricultural science, ecology, botany, physiology, anthropology, and linguistic training.

But the literary impulse has always been present, too. Even as he was completing his Ph.D. dissertation at Arizona, he published his first book of nonfiction ethnobotanical nature writing, *The Desert Smells Like Rain,* in 1982. A year later, he helped to start the nonprofit organization Native Seeds/SEARCH, which aimed to conserve traditional crops, seeds, and farming practices of native people in the American Southwest and northern Mexico. His second book, *Gathering the Desert,* appeared in 1985 and received the John Burroughs Medal for outstanding natural history writing the following year. Even as he developed his reputation as a literary nature writer, Gary simultaneously worked at scientific research and publishing. He has published nearly 100 technical journal articles in fields ranging from geography and nutritional ecology to linguistics and regional studies. He has also published books on environmental topics for general audiences, including *Enduring Seeds: Native American Agriculture and Wild Plant Conservation* (1989) and, with coauthor Stephen L. Buchmann, *The Forgotten Pollinators* (1996). A journey to northern Italy in September 1990 resulted in a foray into travel writing and autobiography, which appeared in 1993 under the title *Songbirds, Truffles, and Wolves: An American Naturalist in Italy.* Also in 1993, he edited a collection of writings called *Counting Sheep: Twenty Ways of Seeing Desert Bighorn Sheep.* Gary then collaborated with Utah nature writer Stephen Trimble on a braided collection of essays on their own childhood experiences and on experiencing the world with their children. *The Geography of Childhood: Why Children Need Wild Places* came out in 1994. *Desert Legends: Re-storying the Sonoran Borderlands,* another collaboration—this one incorporating Gary's essays and distinguished Arizona photographer Mark Klett's visual work—was also published in 1994. The following year he coauthored *Canyons of Color: Utah's Slickrock Wildlands* with Caroline Wilson. Gary's playful, multicultural, transdisciplinary, and engagingly narrative essays were collected again in the 1997 volume

Cultures of Habitat: On Nature, Culture, and Story. His chapbook of poetry, *Creatures of Habitat,* was published in a limited edition in 1998. Recent collaborative projects include *La Vida Norteña: Photographs of Sonora, Mexico* (1998, with T. Sheridan) and *Efrain of the Sonoran Desert: A Lizard's Life among the Seri Indians* (2001, with A. Astorga). *Coming Home to Eat: The Pleasures and Politics of Local Foods* appeared in 2001, and *Singing the Turtles to Sea: The Comcáac (Seri) Art and Science of Reptiles* in 2003.

Although his literary work is most widely known among students, scholars, and aficionados of environmental literature, Gary has famously criticized the restrictive categorization of his own work and that of other "nature writers." As John Elder comments in *Stories in the Land,* "The writer and ethnobotanist Gary Nabhan, resisting the marginalization that he felt in the term 'nature writing,' jokingly remarked before one of his readings that it might be better now simply to refer to books on the same shelf with Thoreau, Leopold, Carson, and Williams as 'literature' and to books on other shelves as 'urban dysfunctional writing.'" Regardless of the phrase one uses to describe Gary's work, it almost inevitably defies single adjectives, typically requiring a string of disciplinary and stylistic descriptors. Perhaps it is best simply to call it literature.

Gary was honored in 1990 with both a Pew Scholarship for Conservation and the Environment and a five-year MacArthur Fellowship. More recently, in 1999, he received a Lannan Literary Award. After spending a number of years as the director of conservation science at the Arizona-Sonora Desert Museum in Tucson, Gary moved to Flagstaff in 2000 to become the founding director of the Center for Sustainable Environments at Northern Arizona University. This is a research center that studies the sustainable use of natural resources on the Colorado Plateau. Gary is married to Laurie Monti. He has two children, Laura Rose and Dustin Corvus, from his first marriage.

My first interaction with Gary Nabhan was merely a visual one. He appeared magically at the 1995 inaugural conference of the Association for the Study of Literature and Environment (ASLE) in Fort Collins, Colorado, a surprise interloper in the ranks of the literati. ASLE was convening on one side of the Colorado State University student union, while the Society for Conservation Biology was holding its major meeting on the other side. Gary had come to Fort Collins to speak to the conservation biologists, but he quickly picked up the literary vibes and began floating, mothlike, between the two conferences, perhaps the only person among the more than 500 scholars to do so. It was during the opening session at the ASLE Conference—featuring talks by William Howarth, SueEllen Campbell, and John Elder—that I first noticed Gary's face, familiar to me from the dustjackets of his books, and I saw him frequently on our side of the building throughout the rest of the weekend.

The serendipitous juxtaposition of ASLE and the Society for Conservation Biology in June 1995 inspired conference participants to suggest that in the future there should be an intentional effort to bring together scholars and artists from various disciplines to discuss our common work on environmental topics. Science and literature, many have argued, should not be artificially separated, but should be brought together in mutually supportive and provocative ways. The kinds of phenomena and experiences we address in our disparate fields are intrinsically extradisciplinary—they require diverse perspectives, a broad range of illuminating discourse. Academic culture and the arts—society in general—require a broad range of interactions to remain vigorous and meaningful. By choosing cross-pollinations as the metaphorical title of this *Credo,* Gary has continued his playful and earnest efforts to inspire activist, scientist, and literary friends to break ranks and seek polyvalent interactions of their own.

SELECTED BIBLIOGRAPHY OF GARY PAUL NABHAN'S WORK

by Ceiridwen McKenzie-Terrill

BOOKS

With Ana Guadalupe Zapata. *Tequila: A Natural and Cultural History.* Tucson: University of Arizona Press, 2004.

Singing the Turtles to Sea: The Comcáac (Seri) Art and Science of Reptiles. Berkeley: University of California Press, 2003.

Coming Home to Eat: The Pleasures and Politics of Local Foods. New York: W. W. Norton and Company, 2002.

With Amalia Astorga. *Efraín of the Sonoran Desert: A Lizard's Life Among the Seri Indians.* El Paso, Tex.: Cinco Puntos Press, 2001.

With Thomas Sheridan and photographs by David Burkhalter. *La Vida Norteña: Photographs of Sonora, Mexico.* Albuquerque: University of New Mexico Press, 1998.

Cultures of Habitat: On Nature, Culture, and Story. Washington, D.C.: Counterpoint Press, 1997.

With Stephen L. Buchmann. *The Forgotten Pollinators.* Washington, D.C.: Island Press, 1996.

With Caroline Wilson. *Canyons of Color: Utah's Slickrock Wildlands.* San Francisco: HarperCollins, 1995.

With photographs by Mark Klett. *Desert Legends: Re-storying the Sonoran Borderlands.* New York: Henry Holt, 1994.

With Stephen Trimble. *The Geography of Childhood: Why Children Need Wild Places.* Boston: Beacon Press, 1994.

Songbirds, Truffles, and Wolves: An American Naturalist in Italy. New York: Pantheon Books, 1993.

With Kevin Dahl. *Conservation of Plant Genetic Resources: Grassroots Efforts in North America*. Nairobi, Kenya: African Centre for Technology Studies Press, 1992.

Enduring Seeds: Native American Agriculture and Wild Plant Conservation. San Francisco: North Point Press, 1989.

With photographs by George H. H. Huey. *Saguaro: A View of Saguaro National Monument and the Tucson Basin*. Tucson: Southwest Parks and Monuments Association, 1986.

With illustrations by Paul Mirocha. *Gathering the Desert*. Tucson: University of Arizona Press, 1985.

The Desert Smells Like Rain: A Naturalist in Papago Indian Country. San Francisco: North Point Press, 1982.

EDITED BOOKS

With John L. Carr. *Ironwood: An Ecological and Culturual Keystone of the Sonoran Desert*. Washington, D.C.: Conservation International, 1994.

Counting Sheep: Twenty Ways of Seeing Desert Bighorn. Tucson: University of Arizona Press, 1993.

With Jane Cole. *Arizona Highways Presents Desert Wildflowers*. Phoenix: Arizona Highways, 1988.

POEMS

"The Seed of a Song" and "The Village on the Other Side of White Horse Pass." In *Getting Over the Color Green: Contemporary Environmental Literature of the Southwest*, edited by Scott Slovic, 225–26. Tucson: University of Arizona Press, 2001.

"Comcáac Indian Horned Lizard Song." *Orion* 20, no. 1 (Winter 2001): 61.

"Coming Out on Solid Ground After the Ice Age," In *Poetry Comes Up Where It Can: Poems from the Amicus Journal, 1990–2000*, edited by Brian Swann, 85. Salt Lake City: University of Utah Press, 2000.

"If the Raven Should Croak Before I Wake." In *New Earth Reader: The Best of Terra Nova,* edited by David Rothenberg and Marta Ulvaeus, 175–76. Cambridge: MIT Press, 1999.

"Seri Indian Horned Lizard Song." *Alaska Quarterly Review* 16, nos. 3 & 4 (Spring/Summer 1998): 171.

Creatures of Habitat: Poems. Berkeley, Calif.: Tangram Press, 1998. Chapbook.

"Sonora Desert Unfolding." In *Spirit Land: Poems by Kim Stafford and Gary Paul Nabhan.* Tempe, Ariz.: Cabbagehead Press, 1997. Limited-edition artists' book (letterpress, shaped handmade paper, multiple-block reduction woodcuts) produced in collaboration by John Risseeuw and Margaret Prentice.

"If the Raven Should Croak Before I Wake." *Terra Nova* 1, no. 2 (Spring 1996): 141.

"Every Night." In *Poetry from the Amicus Journal,* edited by Brian Swann and Peter Borrelli. Palo Alto, Calif.: Tioga Publishing, 1990.

"Diana-Gone-Driftwood Dune Woman." *Great Lakes Review* 3, no. 2 (Summer 1977): 82–89.

"White Horse Pass" and "Through the Night Comes Morning-Talking." *South Dakota Review* 14, no. 4 (Winter 1976-1977): 58–59.

"The Only Desert Spring Evening." *New Mexico Magazine* 54, no. 5 (May 1976): 9.

"Chasing Magdalena." *Blue Cloud Quarterly* 22, no. 4 (1976): 8.

"Diana-Gone-Driftwood Dune Woman." *Kuksu: Journal of Back Country Writing* 5 (1976): 26.

"Lost Pioneer." *South Dakota Review* 11, no. 2 (Summer 1973): 83.

"Bandana Days/The Verde View Unedged." *Tucson Mountain Newsreal* 3, no. 71 (December 1975): 4.

"Those Who Showed Up Joined the Household." *Dacotah Territory* 10 (Spring/Summer 1975): 10–11.

"Spring Smelt Run, Gary Beach, Lake Michigan." *Wisconsin Review* 10, no. 2 (April 1975): 24.

"Tierra Incognita." *High Country News* (February 28, 1975).

ESSAYS, ARTICLES, AND SCIENTIFIC PAPERS

"Nectar Trails of Migratory Pollinators: Restoring Corridors on Private Lands." *Conservation Biology in Practice* 2, no. 1 (Winter 2001): 20–27.

With Joshua J. Tewksbury. "Seed Dispersal: Directed Deterrence by Capsaicin in Chillies." *Nature* 412, no. 6845 (July 26, 2001): 403–4.

"Interspecific Relationships Affecting Endangered Species Recognized by O'odham and Comcáac Cultures." *Ecological Applications* 10, no. 5 (October 2000): 1288–95.

"Desert Ironwood Primer: Biodiversity and Uses Associated with Ancient Legume and Cactus Forests in the Sonoran Desert." Tucson: Arizona-Sonora Desert Museum, 95 pp. (Report), 2000.

With Joshua J. Tewksbury et al. "In Situ Conservation of Wild Chiles and Their Biotic Associates." *Conservation Biology* 13, no. 1 (February 1999): 98–107.

With Gordon Allen-Wardell et al. "Potential Consequences of Pollinator Declines on the Conservation of Biodiversity and Stability of Food Crop Yields." *Conservation Biology* 12, no. 1 (February 1998): 8–17.

"Why Chilies Are Hot." *Natural History* 106, no. 5 (June 1997): 24–29.

"Natural Excursions: Working Under the Influence." *Orion: People and Nature* 15, no. 1 (Winter 1996): 68–69.

With Humberto Suzán and Duncan T. Patten. "The Importance of *Olneya tesota* As a Nurse Plant in the

Sonoran Desert." *Journal of Vegetation Science* 7, no. 5 (October 1996): 635–44.

With Humberto Suzán and Duncan T. Patten. "Nurse Plant and Floral Biology of a Rare Night-Blooming Cereus, *Peniocereus striatus* (Brandefee) F. Buxbaum." *Conservation Biology* 13, no. 6 (December 1999): 461–70.

With Humberto Suzán. "Boundary Effects on Endangered Cacti and Their Nurse Plants in and near a Sonoran Desert Biosphere Reserve." *Ironwood: An Ecological and Cultural Keystone of the Sonoran Desert.* Occasional Papers in Conservation Biology, vol. 1. Washington D.C.: Conservation International, 1994.

With Ted Fleming. "The Conservation of New World Mutualisms." *Conservation Biology* 7, no. 3 (September 1, 1993): 457–59.

"Hummingbirds and Human Aggression: A View from the High Tanks." *Georgia Review* 46, no. 2 (Summer 1992).

"Restoring and Re-Storying the Landscape." *Restoration and Management Notes* 9, no. 1 (Summer 1991): 3–4.

With Donna House et al. "Conservation and Use of Rare Plants by Traditional Cultures of the U.S./Mexico Borderlands." *Biodiversity: Culture, Conservation, and Ecodevelopment.* Edited by Margery Oldfield and Janis Alcorn. Boulder, Colo.: Westview Press, 1991.

With J. C. Brand et al. "Plasma Glucose and Insulin Responses to Traditional Pima Indian Meals." *American Journal of Clinical Nutrition* 51, no. 3 (March 1, 1990): 416–20.

With Eric Mellink et al. "Influences on Habitat and Biotic Diversity: Quitovac Oasis Ethnoecology." *Journal of Ethnobiology* 2, no. 2 (December 1982): 124–43.

With Richard S. Felger. "Ancient Crops for Desert Gardens." *Organic Gardening and Farming* 24, no. 2 (1977): 34–42.

INTERVIEWS

"Voices of Change." Vol. 3. Foundation for Global
Community, 1997. (Foundation for Global Community;
800-707-7932).

"Art of the Wild." Foundation for Global Community, 1996.
(Foundation for Global Community; 800-707-7932).

Goodstein, Carol. *Omni* 16, no. 10 (July 1994): 6.

Satterfield, Terre, and Scott Slovic. "Thriving on Ambiguity:
Latency, Indirection, and Narrative." *What's Nature
Worth? Narrative Expressions of Environmental Values.*
Salt Lake City: University of Utah Press, 2004.

BIOGRAPHICAL/CRITICAL STUDIES AND BOOK REVIEWS

Becher, Anne. "Gary Nabhan." *American Environmental
Leaders: From Colonial Times to the Present.* Santa Barbara:
ABC-CLIO, 2001.

Bernhardt, Peter. Review of *Songbirds, Truffles, and Wolves:
An American Naturalist in Italy. New York Times Book
Review* (July 11, 1993).

Dowie, Mark. Review of *Cultures of Habitat: On Nature,
Culture, and Story. New York Times* (November 30,
1997): 33.

Elder, John. "Teaching at the Edge." *Stories in the Land:
A Place-Based Environmental Education Anthology,* 1–15.
Great Barrington, Mass.: Orion Society, 1998.

Iltis, Hugh H. Review of *Gathering the Desert. Natural History*
95 (March 1986): 94.

Mabberley, David. Review of *Gathering the Desert. Times
Literary Supplement* (May 9, 1986).

Newport, Mary Ellen. Review of *Songbirds, Truffles, and
Wolves: An American Naturalist in Italy. Antioch Review*
52, no. 1 (Winter 1994): 161.

Schueller, Gretel H. "Eat Locally (Think Globally)." *Discover*
22, no. 5 (May 2001): 72–77.

St. Antoine, Sara L. "Gary Paul Nabhan." Vol. 2. *American Nature Writers,* edited by John Elder. New York: Scribner's, 1996.

Terrill, Ceiridwen. Review of *Coming Home to Eat: The Pleasures and Politics of Local Foods. ISLE: Interdisciplinary Studies in Literature and Environment* 9, no. 2 (Summer 2002).

ACKNOWLEDGMENTS FOR "CROSS-POLLINATIONS"

by Gary Paul Nabhan

Thanks to Scott Slovic, Ceiridwen McKenzie-Terrell, Chip Blake, Laurie Buss, and Emilie Buchwald for seeing this project through. Scott Slovic, Alison Deming, Kim Stafford, Karen Gunderson, and Scott Russell Sanders offered valuable feedback on creativity in science and art, and on these essays. As always I'm grateful to my many scientific collaborators, including Steve Buchmann, Humberto Suzán, Rob Robichaux, Laurie Monti, Sara St. Antoine, Josh Tewksbury, Patty West, and Rob Raguso. I am indebted to the John T. and Catherine C. MacArthur Foundation, the Lannan Foundation, and Agnese Haury for fellowships that allowed me to write this manuscript.

WORKS CITED

p. viii   Michael Brenson, *Visionaries and Outcasts* (New York: New Press, 2001), 156.

p. 3   Paul Strand, *The World on My Doorstep: 1950–1976* (Millerton, N.Y.: Aperture, 1994).

p. 10   Gary Paul Nabhan, "Every Night," in *Creatures of Habitat* (Berkeley, Calif.: Tangram Press, 1998), 1.

p. 15   Jakob Bernoulli, *Ars Conjectandi* (1713).

p. 27   T. I. Lubart, "Creativity Across Cultures" in *Handbook of Creativity,* ed. Robert J. Sternberg (Cambridge: Cambridge University Press, 1999), 344.

p. 28   José Louis Brennan, trans., "Pima Jimsonweed Song," in *The Pima Indians* by Frank Russell (Tucson: University of Arizona Press, 1975), 299–300.

p. 41   Albert Szent-Györgyi, in *The Science of Life: A Picture History of Biology* by Gordon Rattray Taylor (London: Thames and Hudson, 1963).

p. 41   James Melchert, in *Visionaries and Outcasts* by Michael Brenson, 158.

p. 45   H. Poincaré, *The Foundations of Science,* trans. George Bruce Halsted (New York: Science Press, 1913), 386.

p. 49   Bentley Glass, in *The End of Science: Facing the Limits of Knowledge in the Twilight of the Scientific Age* by John Horgan (Reading, Mass.: Addison-Wesley, 1996).

p. 52   Jacob Bronowski, *The Ascent of Man* (Boston: Little, Brown and Company, 1974), 153.

p. 52   Amy Clampitt, "Urn-Burial and the Butterfly Migration," in *The Collected Poems of Amy Clampitt* (New York: Alfred A. Knopf, 1997), 132–33.

p. 53   Carl Sandburg, "56. Fog," in *Chicago Poems* (New York: Henry Holt, 1916).

p. 56   Lubart, "Creativity Across Cultures," 344.

p. 59   Peter Forbes, *The Great Remembering* (San Francisco: Trust for Public Land, 2001), 28.

p. 63   Leo P. Kadanoff, *From Order to Chaos II, Essays: Critical, Chaotic, and Otherwise* (River Edge, N.J.: World Scientific, 1999).

p. 67   Alison Hawthorne Deming, "Under the Influence of Ironwoods" (uncollected poem, 2000).

p. 68   Richard Shelton, "Requiem for Sonora," in *Of All the Dirty Words* (Pittsburgh: University of Pittsburgh Press, 1972), 75–76.

p. 68   Nabhan, "Cryptic Cacti on the Borderline," in *Desert Legends* (New York: Henry Holt, 1994), 41.

p. 70   Edward Abbey, *Cactus Country* (New York: Time-Life Books, 1973).

p. 72   Humberto Romero Morales, "El respeto a toda
          criatura viviente" (uncollected poem, 2000).

p. 72   Alberto Búrquez Montijo, "Arbol de la Vida,"
          in *The Ironwood Primer* (Tucson: Arizona-Sonora
          Desert Museum, 2000), 8.

p. 80   Stephen L. Buchmann and Gary Paul Nabhan,
          The Forgotten Pollinators (Washington, D.C.:
          Island Press, 1996), 5.

p. 81   Nabhan, "Hornworm's Homeground: Conserving
          the Diversity of Interactions," in *Wings: Essays
          on Invertebrate Conservation* (Spring 1997): 16,
          17, 19–20.

p. 82   Nabhan, *Cultures of Habitat* (Washington, D.C.:
          Counterpoint, 1997), 254, 255, 257, 258, 259.

p. 84   Scott Russell Sanders, "Beneath the Smooth
          Skin of America," in *Writing from the Center*
          (Bloomington: Indiana University Press, 1995),
          16–17.

p. 85   Nabhan, *Coming Home to Eat* (New York: W. W.
          Norton, 2002), 46.

p. 86   Nabhan, *Coming Home to Eat,* 66.

p. 88   Gretel H. Schueller, "Eating Locally (Thinking
          Globally)," in *Discover* (May 2001): 72, 77.

p. 91   John Elder, "Teaching at the Edge," in *Stories
          in the Land* (Great Barrington, Mass.: Orion
          Society, 1998), 7.

SCOTT SLOVIC, founding president of the Association for the Study of Literature and Environment, currently serves as editor of the journal *ISLE: Interdisciplinary Studies in Literature and Environment.* He is the author of *Seeking Awareness in American Nature Writing: Henry Thoreau, Annie Dillard, Edward Abbey, Wendell Berry, Barry Lopez* (University of Utah Press, 1992); his edited and coedited books include *Getting Over the Color Green: Contemporary Environmental Literature of the Southwest* (University of Arizona Press, 2001), *The Isle Reader: Ecocriticism, 1993–2003* (University of Georgia Press, 2003), and *What's Nature Worth? Narrative Expressions of Environmental Values* (University of Utah Press, 2004). He is professor of English at the University of Nevada, Reno, where he chairs the Literature and Environment Graduate Program.

MORE BOOKS ON THE WORLD AS HOME
FROM MILKWEED EDITIONS

To order books or for more information,
contact Milkweed at (800) 520-6455
or visit our website (www.worldashome.org).

THE *CREDO* SERIES

Brown Dog of the Yaak:
Essays on Art and Activism
Rick Bass

Winter Creek:
One Writer's Natural History
John Daniel

Writing the Sacred into the Real
Alison Hawthorne Deming

The Frog Run:
Words and Wildness in the Vermont Woods
John Elder

Taking Care:
Thoughts on Storytelling and Belief
William Kittredge

An American Child Supreme:
The Education of a Liberation Ecologist
John Nichols

Walking the High Ridge:
Life As Field Trip
Robert Michael Pyle

The Dream of the Marsh Wren:
Writing As Reciprocal Creation
Pattiann Rogers

The Country of Language
Scott Russell Sanders

Shaped by Wind and Water:
Reflections of a Naturalist
Ann Haymond Zwinger

OTHER WORLD AS HOME BOOKS

Toward the Livable City
Edited by Emilie Buchwald

Wild Earth:
Wild Ideas for a World Out of Balance
Edited by Tom Butler

The Book of the Everglades
Edited by Susan Cerulean

Swimming with Giants:
My Encounters with Whales, Dolphins, and Seals
Anne Collet

The Prairie in Her Eyes
Ann Daum

Ecology of a Cracker Childhood
Janisse Ray

Wild Card Quilt:
Taking a Chance on Home
Janisse Ray

Back Under Sail:
Recovering the Spirit of Adventure
Migael Scherer

Of Landscape and Longing:
Finding a Home at the Water's Edge
Carolyn Servid

The Book of the Tongass
Edited by Carolyn Servid and Donald Snow

Homestead
Annick Smith

Testimony:
Writers of the West Speak On Behalf of Utah Wilderness
Compiled by Stephen Trimble and Terry Tempest Williams

THE WORLD AS HOME, the nonfiction publishing program of Milkweed Editions, is dedicated to exploring our relationship to the natural world. Not espousing any particular environmentalist or political agenda, these books are a forum for distinctive literary writing that not only alerts the reader to vital issues but offers personal testimonies to living harmoniously with other species in urban, rural, and wilderness communities.

MILKWEED EDITIONS publishes with the intention of making a humane impact on society, in the belief that literature is a transformative art uniquely able to convey the essential experiences of the human heart and spirit. To that end, Milkweed publishes distinctive voices of literary merit in handsomely designed, visually dynamic books, exploring the ethical, cultural, and esthetic issues that free societies need continually to address. Milkweed Editions is a not-for-profit press.

Join Us

Since its genesis as *Milkweed Chronicle* in 1979, Milkweed has helped hundreds of emerging writers reach their readers. Thanks to the generosity of foundations and of individuals like you, Milkweed Editions is able to continue its nonprofit mission of publishing books chosen on the basis of literary merit—the effect they have on the human heart and spirit—rather than on the basis of how they impact the bottom line. That's a miracle that our readers have made possible.

In addition to purchasing Milkweed books, you can join the growing community of Milkweed supporters. Individual contributions of any amount are both meaningful and welcome. Contact us for a Milkweed catalog or log on to www.milkweed.org and click on "About Milkweed," then "Supporting Milkweed," to find out about our donor program, or simply call 800-520-6455 and ask about becoming one of Milkweed's contributors. As a nonprofit press, Milkweed belongs to you, the community. Milkweed's board, its staff, and especially the authors whose careers you help launch thank you for reading our books and supporting our mission in any way you can.

Typeset in Stone Serif
by Stanton Publication Services
on the Pagewing Digital Publishing System.
Printed on acid-free 55# New Leaf EcoBook
100 percent postconsumer waste recycled paper
by Friesen Corporation.